Number Names

This book is part of the Peter Lang Regional Studies list.
Every volume is peer reviewed and meets
the highest quality standards for content and production.

PETER LANG
New York • Bern • Berlin
Brussels • Vienna • Oxford • Warsaw

Larry J. Schulz

Number Names

The Magic Square Divination of Cai Chen 蔡沈 (1167–1230)

PETER LANG
New York • Bern • Berlin
Brussels • Vienna • Oxford • Warsaw

Library of Congress Cataloging-in-Publication Control Number: 2021038376

Bibliographic information published by **Die Deutsche Nationalbibliothek**.
Die Deutsche Nationalbibliothek lists this publication in the "Deutsche Nationalbibliografie"; detailed bibliographic data are available on the Internet at http://dnb.d-nb.de/.

ISBN 978-1-4331-8599-1 (hardcover)
ISBN 978-1-4331-8684-4 (ebook pdf)
ISBN 978-1-4331-8685-1 (epub)
DOI 10.3726/b18215

Peter Lang Publishing, Inc., New York
80 Broad Street, 5th floor, New York, NY 10004
www.peterlang.com

In memory of our son,
Daniel Howland Schulz
1988-2021
浮生若夢

TABLE OF CONTENTS

List of Figures ix
List of Tables xi
Acknowledgments xiii
Introduction xv

Chapter 1. Numbers 1
A. The Luoshu Magic Square 4
B. Number and Number Sets 6
C. Cyclical Template of 81 Nodes Rendered as Square Matrix 14
D. Structure of the 81 Number Name Matrices 16
E. Structure of the 81 Matrix Suite 19
F. Summary of Rules of Construction for the 81 Matrix Suite 25
G. Architecture and Meaning in the 81 Matrix Suite 28
H. Probabilities of Occurrence Among the Luck Schedule Terms 29
I. Bias Toward Positive Outcomes in 12th-Century Divination Systems 34

Chapter 2. Images 41
A. Chart of the 81 Numbers 45
B. The Language of Cai Chen's Number Names 127
C. The Set of Number Names 128
D. The Agricultural Calendar in the Placement and Diction of Number Names 131
E. Allusions to the *Yijing* and Other Classic Literature 134
F. Qualities and Discipline of the Noble-Minded Person 135
G. Architecture of the Images 138
Conclusion: The Numbers Say . . . 139

Appendix I. *Cai Chen's* Songshi *Biography* 153
Appendix II. *Bibliographic Note* 155
Appendix III. *The "Hongfan School"* 161
Appendix IV. Hongfan *Divination Method* 163
Appendix V. *Glossary of* Yijing *Hexagram and Commentary Names* 171
Appendix VI. *Glossary of Chinese Names and Terms* 173
Selected Bibliography 177
Index 181

FIGURES

Figure 1.1:	Chart of the Eight Numbers in Circular Flow (*bashu zhouliutu* 八数周流图)	9
Figure 1.2:	Basic -3, -3, +1 Replacement Formula in *Luoshu* Terms	12
Figure 1.3:	Nine by Nine Circular Number Graph	14
Figure 1.4:	Nine by Nine Numbers in Rows Chart (left); Conversion to Arabic Numerals (right)	15
Figure 1.5:	Elements of Cai Chen Matrices	17
Figure 1.6:	Interrelationships of the 81 Matrices	29

TABLES

Table 1.1:	*Luoshu* Configuration	4
Table 1.2:	Order 9 Magic Square in Yang Hui, *Xuguzhaiqi suanfa*	5
Table 1.3:	Cammann's Proposed Proto-Order 9 Magic Square	5
Table 1.4:	Luck Schedule Terms in Alpha and Numeric	8
Table 1.5:	Numeric Luck Schedule Matrix	9
Table 1.6:	Luck Schedule Column Replacement Pattern	11
Table 1.7:	Quadrant Detail of Luck Schedule Column Replacement Strategy	12
Table 1.8:	Counting Rod Numerals 1–9	17
Table 1.9:	1.1 Origin Matrix	19
Table 1.10:	1.2 Hiding Matrix	20
Table 1.11:	1.5 Straightness Matrix	21
Table 1.12:	2.1 Achievement Matrix	22
Table 1.13:	2.5 Constancy Matrix	23
Table 1.14:	Matrix Family N, Cell 1.1 Left-Hand Term	23
Table 1.15:	5.1 Abundance Matrix	24
Table 1.16:	5.5 Center Matrix	24
Table 1.17:	Distribution of Terms in 81 Number Name Spectrums	30
Table 1.18:	Positive and Negative Qualities of Luck Schedule Terms	32

Table 1.19:	Frequency of Luck Schedule Terms	33
Table 2.1:	Number Names in Chinese	42
Table 2.2:	Number Names in English	42
Table 2.3:	Antonyms Among Number Names	129
Table 2.4:	Number Names Opposed Across Seasonal Continuum	132
Table A.1:	Aspect Conversion of Heading	167

AKNOWLEDGMENTS

I would like to thank: Xia Shihua 夏世华, the translator-editor of my *"Zhouyi" guaxuwenti zonglun* 〈周易〉卦序问题综论,for providing a survey of contemporary research in Chinese on Cai Chen; mathematician Ryan Hutchinson for offering useful comments on the "Numbers" chapter; and Edward L. Shaughnessy, my touchstone on the early history of the *Yijing*, for making observations that helped shape the "Images" chapter. Thanks to Peter Lang Publishing for bringing this long obscure work to light and to the editorial staff there for their professional assistance in shepherding the book through the publication process. And, as always I am grateful to the love of my life, Barbara, for thoughts on the roughest of drafts and a sympathetic ear to teatime discourses on the idiosyncratic scholars who produced traditional Chinese number theory.

INTRODUCTION

When the NeoConfucian patriarch Zhu Xi 朱熹 (1130–1200) endorsed the set of graphic depictions and divination lore that became the opening chapter in official editions of the *Yijing*,[1] he effectively sanctioned an option for creative expression within what would otherwise become a stifling orthodoxy—an orthodoxy based to a great extent on Zhu Xi's other pronouncements. His *Yijing* graphics include two visualizations of numerical relationships among the first ten digits, called the *Hetu* 河圖 and *Luoshu* 洛書, two arrangements of the eight trigrams (*bagua* 八卦), and several constructs of the 64 hexagrams in square and round arrays.[2] The material also includes Zhu Xi's reconstruction of the yarrow stalk-sorting divination method, and together these graphics and numeric speculations became primary sources for the development of what has been called the "Image and Number School (*xiangshupai* 象數派)."[3]

The new possibilities seen in the graphics Zhuxi selected led to a proliferation of ideas expressed in matrices and charts of various designs that in turn underwrote speculation about relationships among numerical structures and the verbal images of the divination texts in the *Yijing*. So powerful could these relationships be in explaining the *Yijing*'s obscure wording that Image- and Number-type explanations came to be given equal prominence with textual exegesis by the time Lai Zhide (來知德, 1525–1604), the best known of the later Image and Number proponents, wrote his *Yijing* commentary.[4] Lai and other scholars associated with this "school" tended to be independent, and many of them ignored or were dismissive of earlier thinkers;[5] their ideas, though sharing numerous fundamental concepts, did not coalesce into a unified theoretical framework. My earlier research has aimed at pulling together their number theory into a coherent whole, comprising the structure of the sequence in which the 64 hexagrams are presented in the *Yijing* and calendric associations prominent in the Han Dynasty, as well as Zhu Xi's graphics and their successors.[6]

The latter were introduced for the most part by Shao Yong 邵雍 (1011–1077), who is said to have taken some graphic structures from a line of transmission dating back to Chen Tuan 陳摶 (871？ –989), a shadowy character on the cusp between Daoism and Confucianism.[7] Shao also constructed his own depictions, including an 8 x 8 square hexagram matrix surrounded by a related circular array that will play a key role in Part I below. Shao and other Image and Number proponents thought in terms of a circular template betokening the annual solar cycle. They saw the trigrams and hexagrams as numbers marking geometric nodes on that template, which also accommodated calendrical sets of 4 seasons, 12 months (both solar and lunar), 60 "stem and branch" duodecimal counters, 365 days, spatial sets of four or eight directions, the Five Phases (wuxing 五行) of *fengshui* (風水) and medical theory, and other imagistic sets. *Yijing* lore, like strings of associations from the *Shuogua* 說卦, one of the "Ten Wings (*shiyi* 十翼)"—first order commentaries thought to have been written by Kongzi (孔子, Confucius)—could also be located along the template to enrich the correspondences at the nodes. Any member of any set falling on or near the same node could serve as analogs; thus, verbal images such as trigram and hexagram names were taken to reflect the underlying numeric.

From this template, then, one could trace the association of a number—like a trigram or duodecimal counter—with a verbal image. The trigram *zhen* ☳, for example, might occur at the vernal equinox, 90° along the circumference from the winter solstice at 0°. There， also， was located the duodecimal term *mao* 卯, which is thus linked to the trigram *zhen* and all its synonyms: the east, springtime, thunder, action, wood, etc. Of course, the fascination with these depictions, as with the *Yijing* itself, was their assumed efficacy in supporting divination, which involved another complex line of numerical thinking (as discussed in Appendix IV). Thus, divination might be supposed to "work" because its random operation produced a number that intercepted the seasonal template, depicting the current interrelationship of numerical cycles and their unfolding implications in verbal analogs.[8]

I had not, however, come across an Image and Number scholar who expressed such a "hybrid" theory in its entirety. Then I chanced upon the *Master Plan Supreme Pivot Inner Writings* (*Hongfan huangji neipian* 洪範皇極內篇)[9] of Cai Chen 蔡沈 (t. *Zhongmo* 仲默, h. *Jiufeng* 九峰 1167–1230),[10] a first-generation disciple of Zhu Xi. Cai Chen did not work from *Yijing*-related binary figures; instead, he evolved an original system from his own reading of the "*Hongfan*," a chapter in one of the other Confucian Classics, the *Shujing* 書經. This allowed him to build from scratch, drawing on aspects of the *Yijing* tradition—its wording, commentaries, and philosophical associations—while rationalizing many of its anomalies. In this, he followed Shao Yong, whose sprawling *Huangji jingshi shu* 皇極經世書 took *Yijing* binary thinking in unprecedented directions, presenting an intricate model of recurring temporal epochs laid out with reference to the meanings of *Yijing* hexagrams. The "Supreme Pivot (*huangji*)" in the title of Cai's book is in part homage to Shao's invention. What resulted in Cai's case was a set of 81 "Number Names"[11] that ran in sequence around a circular template. Each Name was taken as the function of one of an integrated suite of 81 numerical 9 x 9 matrices, and for each he proposed divination readings (called "legends" herein). These legends are relatively intelligible, in contrast to *Yijing* texts, and Cai also proposed a rationalized yarrow stalk divination method to replace the inordinately complex *Yijing* method as it was understood by Zhu Xi and others. This method yielded in divination a "Number Name (*shuming* 數名)" and a secondary straightforward judgement of its good or bad fortune.

Cai said that his matrices were derived from the *Luoshu* diagram, which he claimed to discover in the wording of the *Hongfan*. The *Luoshu* is recognized by historians of mathematics as the earliest expression of a normal magic square of order 3. Although it appears to have a history that may reach back to the 2nd century BCE., the *Luoshu* was first described in the work of Cai's younger contemporary, the mathematician Yang Hui 楊輝 (d. 1298 CE).[12] Schuyler Cammann argued that in discussing what Yang Hui called "vertical and horizontal charts (*zonghengtu* 縱橫圖), Yang "admitted that he was merely handing down the works of men of old, and he made no claim to being any sort of an innovator himself."[13] Nor did Cammann think Yang Hui fully understood what he was passing on, being on the declining slope of the understanding of magic squares in China: "although the early development of magic squares in China was indeed impressive, judging from the results that have come down to us, this must have reached its height some time before 1275, when Yang Hui published his examples of the early work."[14]

Neither Yang Hui nor Cammann mentions Cai Chen's suite of matrices, nor does Joseph Needham, in *Science and Civilization in China*, or Wu Wenjun in his compendium on Chinese mathematics,[15] but they may well have been the "height" of the art of magic square theory reached "before 1275." With no antecedents of comparable complexity,[16] they emerged suddenly half a century before Yang's book, and, though they attracted some Image and Number enthusiasts who used them to their own ends—a so-called "Hongfan School (*Hongfanpai* 洪範派)," [17] the structure of the matrices was not fully explained either by their author or later scholars. As systematic theoretical work on magic squares only began with 17th-century European mathematicians, Cai's creation has thus remained an unexplored milestone in the history of mathematics—an oversight that the present study seeks to address.

Cai saw "Number (*shu* 數)" as the abstract complement to the *Yijing*'s concrete images (*xiang* 象)—verbal expressions associated with its hexagrams and lines (*yao* 爻). Similarly, his work recasts the *Yijing*'s underlying concepts in an alternate numerical framework and offers what is essentially a commentary on them. The *Yijing*'s images arise, this perspective suggests, from the demarcations that language assigns in expressing the periodicity of natural phenomena. The seasonal round flows from its Origin—the first of his Number Names—at the winter solstice through 81 nodes to reach its End—the Name of the 81st—and immediately, with no interruption, begins again and repeats endlessly. Each Name is expressed numerically by a pair of the digits one to nine just as each *Yijing* hexagram is taken to reflect the interplay of two trigrams. The latter idea was superimposed on the original *Yijing* text by the first order commentaries; in Cai's system, it is fully integrated from the outset.

Cai explicitly employs the circular template's seasonal analogy as his transition from pure number to verbal imagery. The template is "Pattern" (*li* 理, often translated "Principle"), the signature NeoConfucian philosophical concept.[18] Cai said, "Number is the temporal order of Pattern; words are the meaning of Number."[19] In itself unseen, Pattern is manifest and comprehensible in the life cycles of all phenomena, collectively and individually. Cai takes the numerical representation of Pattern from its circular archetype to a square by transposing its sequential nonary segments to the columns of a 9 x 9 matrix. He then explores the relationships exposed by this juxtaposition as Shao Yong did in his 8x8 hexagram matrix.[20] Cai presented 81 successive permutations of the original graphic that display a unique numerical structure for each Number Name node. His Number Names are the fundamental "words" that are "the meanings of Number" at each node, and Cai also expanded on them in the legends attached to each Name. However, he insisted that all numbers are in effect two numbers, the odd and even of *yang* 陽 and *yin* 陰, which represented the universal cyclical dynamic of extension and intension. In this way, he kept his base nine approach in synchronization with the binary basis of all Image and Number thinking.

In what follows, I detail the architecture of the suite of 81 matrices and the corresponding literary structure of the Number Names and their legends. The former is precise in its logic and adherence to rules; the latter is poetic, relying on analogy and classical allusion. However, analysis of Cai's diction and the placement of terms in their own matrices demonstrates that he employed concepts reflecting his numerical rules in choosing his terms. I also explain how his thinking about the divination process influenced and was influenced by his number theory. Hopefully, the discussion of these themes will help place Cai's work in intellectual history and in the process shed more light on the innovations that the Image and Number School has contributed to the history of mathematics.

Notes

1. For example, in Hu Guang, et al., *Zhouyi zhuanyi daquan*, the official *Yijing* text for civil service examination purposes in the Ming Dynasty. In his *Discussion of the Image and Number in the Study of the Yijing* (*Yixue xiangshu lun*), Huang Zongxi 黄宗昔 (1610–1695) amassed historical proofs to demonstrate that these graphics were extraneous to the *Yijing* per se and were of Daoist provenance. See Huang Zongxi, *Yixue xiangshu lun*, ch.1, *Sikuquanshu tiyao* 四庫全書提要, 1a.
2. These can be found, for example, in Dong Kai, *Zhouyi zhuanyi fulu*, ch. *shouxia* 卷首下. Regarding the binary number theory implicit in trigram and hexagram arrays, see Schulz, *Hexagrammatics* and *Ngua Theory*.
3. The terms 象and數 are used together with reference to divination in the 4th century BCE *Zuozhuan* 左傳 for the 15th year of Duke Xi 僖公十五年: 龜象也；筮數也。物生而後有象；象而後有滋；滋而後有數。The turtle [cracks] are images; yarrow stalk [sortilege] is number. Things are born and afterward have images; after there are images, there is life; after there is life, there are numbers." Kang Yu describes three "waves" of Image and Number practice based on Han Dynasty models of the first wave. The second featured the incorporation of astronomical elements brought into play by the Tang Dynasty monk Yi Xing 一行, and Shao Yong 邵雍 began the third. See Kang Yu, "*Lun songyuan xiangshusi*, 93–99. This "third wave," in which the subject of this study is placed, is sketched in Schulz, *Lai Chih-te*, 10 ff.
4. See Schulz, *Lai Chih-te*, 44 ff. for Huang Zongxi's critique of Image and Number commentary in the late 17th century and 251 ff. for a publishing history of Lai's *Zhouyi jizhu* 周易集注. At least five more editions of Lai's book have been published in China since 2000.
5. See, for example, Schulz, *Lai Zhi-te*, 44 ff.

6. Schulz, "Structural Elements," *Ngua Theory*, and "*Zhouyi*" *guaxuwenti zonglun*.
7. See Huang Zongxi, ch. 1, paragraph 31.
8. See Schulz, *NGua Theory*.
9. The bibliographic note from the *Sikuquansh zongmutiyao* 四库全书总目提要 is translated in Appendix II. The contemporary pronunciation of the character 沈, Cai's given name, might be rendered either *chen*, meaning "deep," or *shen*, which means "juice" or is a surname. His brothers' given names were *Yuan* 渊, "abyss," and *Hang* 沆, "vast expanse of water." These share the water radical with 沈, perhaps making the three pertain to water as one of the Five Phases (*wuxing*). If they are an intentional triplet, the meaning "deep" that goes with the pronunciation *Chen* would be consistent with that of *Yuan* and *Hang*.
10. Cai's biography is appended to his father's in the *Songshi*, 宋史 列傳193. A translation appears in Appendix I.
11. Cai called his general chart in which 81 sequential numeric figures are shown with individual names "General List of the Names of the Numbers in the Supreme Pivot Inner Chapters (*Huangji neipian shuzongming* 皇極內篇數總名)", from which I have taken the term Number Name. Cai Chen, *Huangji*, 3:1a.
12. The *Xuguzhaiqi suanfa* 續古摘奇算法. See Schuyler Cammann, "The Evolution of Magic Squares in China." Joseph Needham cites the *Zhouyi qianzaodu* 周易乾鑿度 (1st or 2nd cen. CE according to a citation in the *Gu weishu* 古魏書) as the earliest concrete reference to the numbers in what was later called the *Luoshu*. *Science and Civilization in China*, vol. 3, 58. Since Needham's book and Cammann's article were published, an astrological compass known as *Taiyi jiuguan zhanpan* 太乙九宫占盘 and bearing the eight perimeter numbers of what came to be know as the *Luoshu* was unearthed from the 173 BCE. tomb of Ruyin Hou 汝阴侯 in Shuanggudui, Fuyang County, Anhui 安徽阜阳县双古堆 in 1977. See https://zh.wikipedia.org/wiki/九宫图.
13. Cammann, "Evolution of Magic Squares," 120.
14. *Ibid.*, 122.
15. Wu Wenjun's *Zhongguo shuxueshi daji* 中国数学史大系 discusses the history of magic squares in China in vol. 5, 622 ff. and 689 ff.
16. Cammann, "Evolution of Magic Squares," 121 describes the lone example of an order 9 magic square among Yang Hui's depictions as "one of the easiest magic squares to make, as it is only a logical extension of the simple square of three." Jacques Sesiano has noted relatively sophisticated Indian antecedents to 9 x 9 magic square construction theory three centuries before Cai Chen's time in *Magic Squares in the Tenth Century*. There is no reason to believe Cai Chen was influenced by external sources, however.
17. So called by the *Sikuquanshu* editors. In Cai Chen, *Hongfan*, *Sikuquanshu tiyao* 4a. See Appendix III.
18. "NeoConfucianism" as it evolved in the Song 宋 Dynasty is known as *lixue* 理學 in Chinese. See Wint-tsit Chan, *A Sourcebook in Chinese Philosophy*, 751.
19. "數者理之時也辭者數之義也." Cai Chen, *Hongfan Neipian*, 2:21a.
20. See Schulz, *Hexagrammatics*. Cai referenced Shao's graphic by adapting it to a 9 x 9 format as his "*jiujiufangshu tu* 九九方數圖," Cai, *Hungfan* 1:2a.

· 1 ·

NUMBERS

Shu

數

Cai Chen is remembered for writing two books at the behest of others. His master, Zhu Xi, had produced works on other Confucian texts but had not in his late years written an intended commentary on the classic *Book [of History]* (*Shujing*), and he commissioned Cai to write it for him. Cai's book, titled *Collected Commentaries on the Book [of History]* (*Shu jizhuan*書集傳), became the orthodox standard text for the civil service examinations from1315 CE. [1] A chapter in the *Shijing* is called "*Hongfan*," and it presents an idealized "Master Plan" for governing the country delivered to King Wu (Wuwang 武王, 11th cen. BCE.) by Jizi 箕子, a courtier of the Shang (尚) Dynasty that King Wu's Zhou (周) Dynasty defeated. Cai Chen's father, Cai Yuanding 蔡元定 (1135–1198), believed there was evidence for a novel number theory in the *Hongfan* and imagined a book on the subject but was unable to complete it. He said, "The one who will complete my book is Chen."[2] This became Cai Chen's second book, *Master Plan Supreme Pivot Inner Writings* (*Hongfan Huangji Neipian*). In Cai's book, divination is the handmaid of philosophical and numerical speculation, as it is in the understanding of the *Yijing* that coalesced under Zhu Xi's amalgamation of Shao Yong's number theory and Cheng yi's 程颐 (1033–1107) textual interpretations.[3] Cai Chen developed a system that integrated both those interests and presented them in an elegant visual format derived from the diagram known as the *Luoshu*, literally "Luo [River] Book."[4]

Cai inherited from Zhu Xi and his father the consensus NeoConfucian ontology. He understood phenomenal existence as a natural process described in the term "that which creates and transforms (*zaohua* 造化)," in which phenomena individually and in concert follow a standard cycle of extension and intension (*qushen* 屈信).[5] The cycles of phenomenal activity, like day and night, summer and winter, complete and begin again without interruption.[6] These cycles are structured according to Pattern(s) (*li*)—one general, overarching Pattern and one individualized and expressed in each of the "myriad things (*wanwu* 萬物)." Pattern(s) are invisible[7] but can be quantified according to their numerical regularity when they take on materiality (*qi* 氣) in identifiable forms (*xing* 形).[8] Cai's Master Plan system is a model of this unitary Pattern employing numerical sets from classical sources and of his own invention. As would be expected, he regularly applies the standard vocabulary of *yin* and *yang*, the Five Phases (*wuxing*), and the 10 heavenly stems (*tiangan* 天干) and 12 earthly branches (*dizhi* 地支). However, the *Luoshu* set of nine numbers, identified with the *Hongfan's* Nine Divisions (*jiuchou* 九疇), is at the core of what is called herein his "Luck Schedule" and the 81 Number Name matrices. It also figures into his divination calculations, which otherwise employ the "numbers of the great expansion (*dayanzhishu* 大衍之數)" and its subdivisions found in the first order *Xici* commentary to the *Yijing*.[9] Also from *Yijing* first order commentaries, the set of eight trigrams represents the informed and verbalized images (*xiang*象) quantified by the *Luoshu*'s 9 numbers.[10]

There were several antecedents to Cai's system in the Image and Number tradition. Meng Xi 孟喜 (1st cen. BCE.) and his disciple Jing Fang 京房 (77–33 BCE.) proposed a method for indexing the 64 hexagrams to the annual solar cycle in what is known as the *guaqitu* 卦氣圖.[11] As 64 does not divide comfortably into 365.25 days, the *guaqi* methodology uses 60 of the hexagrams (assigning the remaining four one each to the four seasons) and makes each of the other hexagrams governor of a 6 7/80 day period. The number 81—which has its own allure as the square of nine, the highest digit among base-ten numerals—divides evenly into 365, yielding 4.5. Thus, a system that had 81 divisions offered sharper definition to spreading divination readings over the annual round than the Meng Xi-Jing Fang approach, and Yang Xiong 楊雄 (53 BCE.–18 CE) turned to that number in his *Taixuanjing* 太玄经 with its 81 tetragrams.[12] In Yang's day, the number 81 had also proven pivotal in fixing a more precise length for the year and aligning the solar and lunar calendars.

When the first "modern" calendar, the *taichuli* 太初曆, was put into practice in 104 BCE., it employed a method for fixing the average length of a month at 29 43/81 days. The denominator of 81 in that fraction was derived from the measurement of the pitch pipe for the fundamental tone *huangzhong* 黄鐘 (middle C). The *Hanshu* says of this, "Their method of constructing the calendar drew on the tonal [ratios]. They said, 'As the volume of the tonal [ratio] in a pitch pipe adds up to 81 *cun*, then that became the divisor for the [remainder] day.'"[13] Both the calendar and the production of pure tones for use in ritual ceremonies were deemed of great importance to successful governance, and numerical similarity between two seemingly unrelated natural phenomena was both uncanny and effective in application. The resulting method of adjusting the numerical irregularities of the year was thus called the "tonal ratio calendar (*lülifa* 履歷法)."

However, the derivation of 81 in that passage was subject to numerous revisions in later works as well as uncertainty as to the length intended by the measurement *cun*, which was variously defined under different dynasties. Cai Chen's father, Cai Yuanding, wrote a study of the mathematics of the tonal ratios called *The NewBook of Pitch Pipes* (*Lülü xinshu* 律吕新書), and some of his ideas on the relationship between numbers and their manifestations in form—in that case, the expression of breath in the pitch of a bamboo tube—carried over into Cai Chen's thinking.

Cai's 81 Number Names are distributed evenly over the solar year, as were the hexagrams in the *Guaqitu* and tetragrams in the *Taixuanjing*. Cai was surely familiar with these,[14] and he acknowledged awareness of two earlier fabricated divination systems the *Dongji zhenjing* 洞極真經 attributed to Guan Lang 關朗 （+5th cen.?) and the *Qianxu* 潜虚of Sima Guang 司馬光 (1019–1086). In his preface, Cai wrote: "The *Dongji* has [aspects of] the [Luo] shu; the *Qianxu* has [aspects of] the [He]tu—these were worth producing."[15] And, as had Shao Yong, Cai expressed his model in an exhaustive numerical framework, also presenting circular and square matrices to illustrate his theory.[16] His father, Cai Yuanding, wrote an extensive critique of Shao's *Huangji jingshi*;[17] so between his teachers and his own reading, he had recourse to all the standard sources for the Image and Number School. However, Cai's use of Image and Number precedents was unconventional and his number theory entirely unique. In this regard, Kang Yu suggests that he reversed the emphasis of Shao and his followers, giving number primacy over image: "The contribution of the Cais, father and son, lies in fundamentally transforming Image and Number Study from divination to a schematic construct of the world;

moreover, they revealed the use of 'number' in the interrelationships among the myriad things."[18]

A. The *Luoshu* Magic Square

"The late noble-minded person [my father,] said, 'The *Luoshu* is the origin of number,'"[19] wrote Cai Chen. The *Luoshu* is a magic square—indeed, is taken to be the oldest example of a "normal" magic square of order three.[20] It has two rules of construction: first, the rows, columns, and diagonals sum to the same number, 15 (i.e., its magic constant (sum of each column and row) M = 15).

Table 1.1: *Luoshu* Configuration

4	9	2
3	5	7
8	1	6

In addition, the numbers are "associative," meaning that if symmetrically equidistant from the center, they sum to the same number: one and nine, two and eight, three and seven, and four and six all sum to ten. This associative property might be called a "tens complement" rule with respect to the *Luoshu*.

In his commentary on the title of the *Shujing's Hongfan* chapter, Cai Shen wrote: "The *Hanzhi* says that when Yu controlled the flood waters, he was given the *Luoshu* concept and by it arranged his [hydraulic works]. That is the 'Master Plan (*Hongfan*).'"[21] As the *Sikuquanshu* editors point out in their bibliographic note to Cai's book, this was an entirely unprecedented, and, in their view, fallacious, reading of the classic text.[22] However, Cai identified the nine numbers in the *Luoshu* with the *Hongfan's* Nine Divisions (*jiuchou*),[23] and Cai took this as his rationale for employing in his work a nonary system, lacking zero and using digits 1–9. He then expanded several structural implications of the order 3 magic square—the centrality of the number five, the ten-complement relationship of the nine perimeter numbers, and the distances between those numbers—into an integrated suite of 81 9 x 9 matrices with the properties of associative and semi-magic squares (in which rows and columns sum to the same number but not diagonals). As discussed below, the matrices transform sequentially based on a formula derived from the *Luoshu*.

Cammann points out that the magic squares in Yang Hui's *Xuguzhaiqi suanfa* were all more or less constructed on the *Luoshu* model; indeed, he thought

Chinese mathematicians "were apparently so hypnotized by this early solution, and by all the cosmic and magical properties ascribed to it, that they continued to try to adapt the [*Luoshu*] principal to the solving of higher squares as well."[24] He reproduces Yang's 9 x 9 magic square (Table 1.2), which showed, he said, "the greatest development of the [*Luoshu*] principle" among Yang's examples.

Table 1.2: Order 9 Magic Square in Yang Hui, *Xuguzhaiqi suanfa*[25]

31	76	13	36	81	18	29	74	11
22	40	58	27	45	63	20	38	56
67	4	49	72	9	54	65	2	47
30	75	12	32	77	14	34	79	16
21	39	57	23	41	59	25	43	61
66	3	48	68	5	50	70	7	52
35	80	17	28	73	10	33	78	15
26	44	62	19	37	55	24	42	60
71	8	53	64	1	46	69	6	51

Cammann said,

"Apparently, the man who invented it first wrote down the sequence of numbers from 1 to 81 in nine parallel columns, after which he took each horizontal line of nine numbers, in turn, and set them out in the order of their sequence, to make from each row a small square on the [*Luoshu*] pattern. Finally, he took these nine subsidiary squares in order and arranged them to construct a much larger square, again on the [*Luoshu*] pattern. The result was a 'composite magic square,' consisting of nine smaller magic squares, each capable of functioning separately".[26]

Table 1.3: Cammann's Proposed Proto-Order 9 Magic Square

1	10	19	28	37	46	55	64	73
2	11	20	29	38	47	56	65	74
3	12	21	30	39	48	57	66	75
4	13	22	31	40	49	58	67	76
5	14	23	32	41	50	59	68	77
6	15	24	33	42	51	60	69	78
7	16	25	34	43	52	61	70	79
8	17	26	35	44	53	62	71	80
9	18	27	36	45	54	63	72	81

This, then, was the state of the art in 1275, and, as will be seen, it was simplistic relative to Cai Chen's matrices. Cammann's explanatory illustration (Table 1.3) showing the sequence 1–81 in vertical columns appears as the "Nine by Nine Numbers in Rows Chart (*jiujiu hangshu tu*)" in Cai's book (reproduced below, Fig. 1.4)). Cai later populated this basic matrix, which he showed running from right to left, with his Number Names and used it as an index for all his 81 matrices. He evolved his matrices in a radically different fashion from that described by Cammann; however, he was true to the *Luoshu* roots of the magic square in other ways we shall discuss. Suffice it to say that the product of his work was sophisticated beyond the scope of Yang Hui's collection. Yet it has been overlooked by historians of mathematics, perhaps because Cai's matrices must be converted from verbal terms to numbers in order to be recognized as such and analyzed.

Cai Chen linked his matrices to 81 Number Names (*shuming*) and to 9 luck terms—verbal expressions that he applied to a unique system of divination.[27] The Ming Dynasty scholar Xiong Zongli 熊宗立 (1409—1482) wrote the definitive commentary on Cai's book, preserved in its *Sikuquanshu* edition. It is from Xiong's glosses that parts of Cai's stalk sorting methodology can be understood and some of his sources in the classical tradition parsed. However, Xiong gave no clue to unravelling the rule-based logic of Cai's matrix construction. Later members of the *Hongfan* School (*Hongfanpai*) tended to carry Cai's ideas into individualistic offshoots rather than elucidating them in their own right. In the Ming Dynasty, for example, Cheng Zongshun 程宗舜 (fl. ca. 1550) expanded Cai's Number Names and legends into an imitation *Yijing* that was detached from Cai's matrices and employed an alternate numerical scheme.[28] Modern research, such as that of Kang Yu, has focused on the philosophical implications of Cai's work and his place among the Image and Number thinkers, leaving the logic of his number system to be deduced from Cai's matrices in the study that follows.

B. Number and Number Sets

In Cai's system, "number" refers to repeating numerical continuums as well as to individual members of those continuums. Such continuums are meaningful in that their members can be understood as nodes of a developmental cycle shared by phenomena in general. The most elemental continuum has two members, 1 and 2, which are *yang* and *yin*, respectively odd (*qi* 奇) and even (*ou* 偶), and concisely describe the fundamental cycle of "extension and intension." Interestingly, the terms 偶 and 耦, both pronounced *ou*[3] in Mandarin, are used with different connotations in Cai's text and add a contextual layer to his meaning. The first character means "even" of numbers, as in "一者，奇；二者，偶也。 One is odd; two is even."[29] The second character's primary meaning is "pair," derived from the idea of plowing side by side in a team. But in the *Yijing Xici* commentary, it is taken to mean "even" with respect to the typology of trigrams: "陽卦奇，陰卦耦 A *yang* gua is odd, a *yin* gua is even."[30] So in "兩耦為二 Two evens make a 2,"[31] Cai's connotation is numerical. But in "耦者，陰陽之對待乎 'Even' is the paired opposition of *yin* and *yang*, is it not,"[32] he defines the term more broadly as the pairing of those two primal states of phenomenal being.

Cai plays on this ambiguity of "even" and "opposition" to express the relationship between the numbers that he believes run invisibly in the background of existence and produce observable phenomena. "If there is no first, then a second cannot be completed; if there is not a second, then the first cannot be realized. The second can be known, and the first is difficult to know [by itself]. The second can be seen and the first is difficult to see. What can be known and seen—is it not substantial? What is difficult to know and to see—is it not the infinitesimal?"[33] Thus, 1 and 2 are counters as well as metaphors for the structure and manifestation of phenomenal images: "Number starts with one and is complete at two. One is odd; two is even. Odd is that whereby number works; even is that by which images are established."[34]

At this fundamental binary level of number, Cai expands on the *yang* and *yin* concepts of traditional philosophy and their dualistic counterparts throughout its lexicon: light and dark, summer and winter, male and female, etc. Here, at the same time that they are the numbers one and two, they are (1) the underlying structural dynamic that drives phenomenal existence, which is invisible and in itself unintelligible; and (2) the types of individual phenomena in their observable life cycles. Together, one and two represent in microcosm the opposition inherent in phenomena and the pattern of growth and decay that opposition occasions. At the same time, they are ancestral to other numerical sets that give finer definition to the unitary cycle: four (seasons *shi* 時), five (Phases), eight (*Yijing* trigrams), ten (heavenly stems), twelve (earthly branches), and 64 (*Yijing* hexagrams).

However, Cai uniquely folded these traditional sets into a number system built upon the nine members of the *Luoshu* in their magic square relationships. The salient features of the *Luoshu* normal magic square are that the number 5 occupies the central place and that it is integral to four of the eight sums of 15 contained in the square. Cai developed his base-nine numeric system around this mediating role of the central five within the nine numbers around the perimeter. He wrote: "Numbers begin with one, are tripled at three, consummated at nine, complete at 81, and fully developed at 6,561."[35] We will return to the role of one and three below in discussing the replacement strategy that allowed Cai's 81 matrices to maintain consistent rules of construction. The number nine guided the development of what is named herein a "Luck Schedule" (Table 1.4) of nine terms that were arrayed into nine different columnar arrangements giving the 81 cells of each matrix and 81 matrices in all. The 81 matrices have a total, then, of 6,561 cells, each containing a pair of Luck Schedule terms.

Cai lays out the terms of the Luck Schedule in pairs that sum to 10 saying, "One is Good Luck and nine is Bad Luck; three is Blessing and seven is Disaster; eight is Felicity[36] and two is Error; four is Remorse and six is Regret. These eight numbers flowing in a circuit extrude types and seek five in the center."[37]

Table 1.4: Luck Schedule Terms in Alpha and Numeric

1	A	吉	Good Luck
2	B	咎	Error
3	C	祥	Blessing
4	D	吝	Remorse
5	E	平	Balance
6	F	悔	Regret
7	G	災	Disaster
8	H	休	Felicity
9	I	凶	Bad Luck

These terms are displayed in a 9 x 9 matrix in which the initial ordinal sequence is reflected in the right-hand column when the luck schedule terms are converted to numbers.[38]

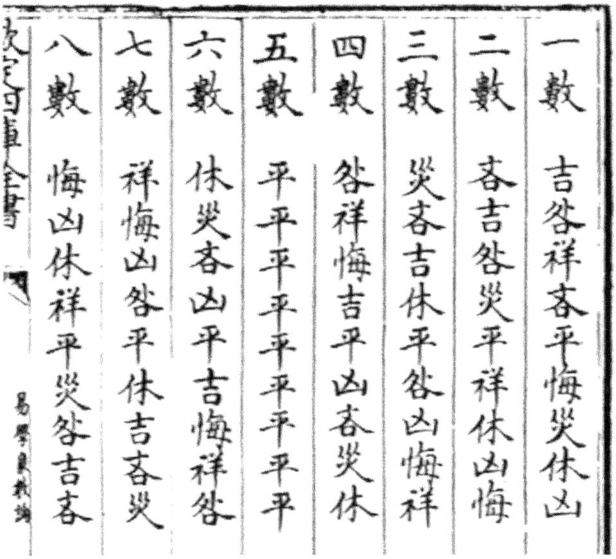
欽定四庫全書

一數 吉咎祥吝平悔災休凶
二數 吝吉咎災平祥休凶悔
三數 災吝吉休平咎凶悔祥
四數 咎祥悔吉平凶吝災休
五數 平平平平平平平平平
六數 休災吝凶平吉悔祥咎
七數 祥悔凶咎平休吉吝災
八數 悔凶休祥平災咎吉吝

Figure 1.1: Chart of the Eight Numbers in Circular Flow (*bashu zhouliutu* 八数周流图)

The "Eight Numbers" in the title of Figure 1.1 refers to the eight numbers one to four and six to nine. Presumably, this discounting of five makes reference to its central position among the *Luoshu* numbers, a quality that makes it transitional and sets it apart from the other numbers.

Translating the terms to Cai's numerical counterparts yields this Luck Schedule matrix:

Table 1.5: Numeric Luck Schedule Matrix, derived from Figure 1.1

9九數	8八數	7七數	6六數	5五數	4四數	3三數	2二數	1一數
9	6	3	8	5	2	7	4	1
8	9	6	7	5	3	4	1	2
7	8	9	4	5	6	1	2	3
6	3	2	9	5	1	8	7	4
5	5	5	5	5	5	5	5	5
4	7	8	1	5	9	2	3	6
3	2	1	6	5	4	9	8	7
2	1	4	3	5	7	6	9	8
1	4	7	2	5	8	3	6	9

This depiction reveals that the "Chart of the Eight Numbers" is a semi-magic square with the magic constant (sum of each column and row) M = 45. Apparent in this depiction are: (1) a tens complement rule and (2) an odd-even placement rule. In addition, the arrangement features: (3) exceptional treatment of the number five, which occupies the entirety of the fifth row and column and appears nowhere else, and (4) the occupation by one and nine of all cells in the two diagonals.

With respect to 1), the tens complement rule is evident in the pairs of numbers radiating outward from the axes of fives. One and nine, two and eight, three and seven, and four and six are in opposing positions equidistant from the horizontal axis in column one on the right and all pairs sum to ten. One and nine, four and six, seven and three, and two and eight oppose each other and sum to ten in the top row. Other numerical orders emerge in the remaining rows and columns, but in all cases a number has a complement that sums to ten symmetrically across the horizontal and vertical axes. (The complement rule only trivially applies to the five axes, where five appears in all cells of its row and column.) Thus, in magic square terminology, rows and columns are all associative internally. Concerning 2), odd and even placement follows a consistent rule for all numbers but five. Otherwise, in odd-numbered columns, odd numbers occupy cells in odd-numbered rows and evens occupy even-numbered rows. In even-numbered columns, even numbers are placed in odd-numbered rows and odds in even-numbered rows.

The Wikipedia entry "Magic Squares" notes that "completely general methods for producing all the magic squares of all orders do not exist." That is, no algorithm has been devised that entirely explains their construction: instead, they are formed number by number using various strategies that involve spatial moves within or wrapping around their grid. For example, in what appears to be the earliest analytical study of the subject, William Symes Andrews describes a "knight move" method of construction that takes the analogy of the chess piece's range of two cells forward and one to either right or left.[39]

The numbers in column 1 proceed top to bottom in ordinal sequence. Rows, however, follow a different logic arising from a column-to-column replacement strategy shown by the arrows in Table 1.6 below. Three types of replacement can be observed in this Table. Columns one and two both follow the first type, in which the first two numbers move down one place and the last two move up one. The third number and the 6th number move down three, and the fourth and seventh move up three. The one-cell movements establish a trend that places one and nine on the matrix's diagonals and along with the two three-cell movements preserves the pairings that sum to ten complements across the horizontal 5 axis.

9九數	8八數	7七數	6六數	5五數	4四數	3三數	2二數	1一數
9	6	3	8	5	2	7	4	1
8	9	6	7	5	3	4	1	2
7	8	9	4	5	6	1	2	3
6	3	2	9	5	1	8	7	4
5	5	5	5	5	5	5	5	5
4	7	8	1	5	9	2	3	6
3	2	1	6	5	4	9	8	7
2	1	4	3	5	7	6	9	8
1	4	7	2	5	8	3	6	9

Table 1.6: Luck Schedule Column Replacement Pattern

Though neither Cai nor any later commentator explains his strategy, his decision to change positions by three or one interval appears to be grounded in the placement of numbers in the *Luoshu* (Figure 1.2). The illustration below shows how the first four terms in each of the first four columns of the Luck Schedule 9 x 9 matrix map to the perimeter numbers of the Luoshu. (Because of Cai's tens complement rule, the rule applied in these four samples forces it to hold for the other quadrants of the matrix, though the arrows would reverse in direction below the horizontal axis.)

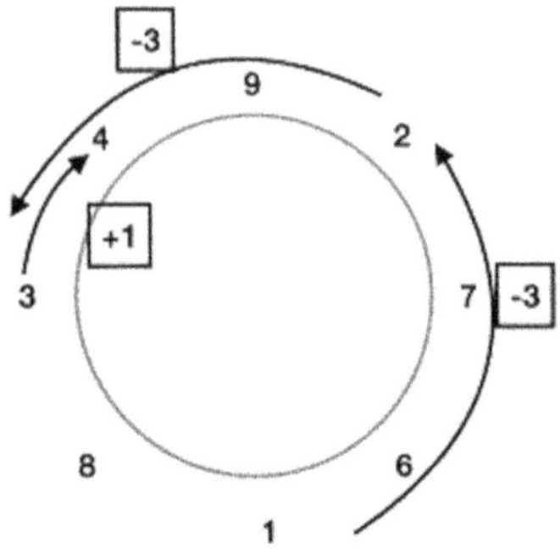

Figure 1.2: Basic -3, -3, +1 Replacement Formula in *Luoshu* Terms

		Col. 1					Col. 3		
col.→	3	2	1	row↓	col.→	3	2	1	row↓
	4D	9I	2B	0.1		4D	9I	2B	0.1
	3C		7G	0.2		3C		7G	0.2
	8H	1A	6F	0.3		8H	1A	6F	0.3
term	A	B	C	D	term	G	D	A	H
col.row	2.3	1.1	3.2	3.1	col.row	3.1	2.3	1.1	1.2
places		-3	-3	1	places		-3	-3	1
		Col. 2					Col. 4		
col.→	3	2	1	row↓	col.→	3	2	1	row↓
	4D	9I	2B	0.1		4D	9I	2B	0.1
	3C		7G	0.2		3C		7G	0.2
	8H	1A	6F	0.3		8H	1A	6F	0.3
term	D	A	B	G	term	B	C	F	A
col.row	3.1	2.3	1.1	1.2	col.row	3.1	2.3	1.1	1.2
places		-3	-3	1	places		-3	-3	1
						(if H)	1	-3	1

Table 1.7: Quadrant Detail of Luck Schedule Column Replacement Strategy

In all three columns, the terms follow the same pattern: the second term is 3 cells counterclockwise from the first (-3), the third is three cells counterclockwise from the second (-3), and the fourth is one cell clockwise from the third (+1).

This -3, -3, +1 pattern is also observable in the first four numbers of Cai's circular depiction of the *Luoshu* titled "Chart of Eight Numbers in Opposition (*Bashuxiangduitu* 八數相對圖),"[40] from which Table 1.7 is taken.

In the quadrant matrices, column 2 of the Luck Schedule begins at 4D (alpha-numeric here to distinguish from columns and rows), where column 1 ends; column 3 begins at 7G, where column 2 ends. Column 3 ends at 8H, which is one of two possible numbers to head column 4. But using 8H in column 4 row 1 would require a clockwise move of one cell from 8H to 3C. The resulting pattern would be +1, -3, -1 as opposed to the consistent -3, -3, +1 achieved by starting at 2B instead. Such an application of *Luoshu* relationships is consistent with Cai's maintenance of the centrality of the number five and imposition of the 10 complement rule in his matrices.

Another effect of placing two rather than eight at the head of the second column in the Luck Schedule matrix is that it clusters the even-numbered "bad luck" members of the spectrum with the generic "Good Luck" heading of 1A in column 1 (and, reciprocally, the three specific types of "good luck" with the "Bad Luck" heading of 9I on the other side of the vertical 5 axis). This would have been a satisfying juxtaposition philosophically whether or not the *Luoshu* pattern were used as a structural factor.

In following the *Luoshu* -3, -3, +1 pattern to the fourth column, 1 and 9 both move one cell further along their diagonals, until they abut the vertical 5 axis. But the effect on the other numbers in the matrix column 4 is expressed in moves of seven cells for the numbers 7 and 3 and five cells for 8 and 2, 6 and 4 in the fourth column. This twist of six of the eight numbers prefigures the subsequent step after the vertical axis, in which column 4 is upended in its entirety to form column 6. The remaining column-to-column replacements in columns 6 through 9 reverse the process described for columns 1–4, making every remaining column the invert of the column at the same distance from the vertical 5 axis.

In these ways, Cai creates a dynamic model of development over time in which the same eight elements are rearranged around the central axes in keeping with simple rules that force later outcomes and reflect what he saw as inherent qualities of numbers—complementariness and oddness or evenness.

C. Cyclical Template of 81 Nodes Rendered as Square Matrix

Cai's book shows the sequence of 81 first as a "Nine by Nine Circular Number Graph (*jiujiu yuanshutu*, 九九圓數圖),"[41] where he treats each degree on the circle as a bipartite number 一一 (transcribed as 1.1), 一二 (1.2), 一三(1.3), . . . 一九(1.9), 二一(2.1), . . .九九(9.9).[42]

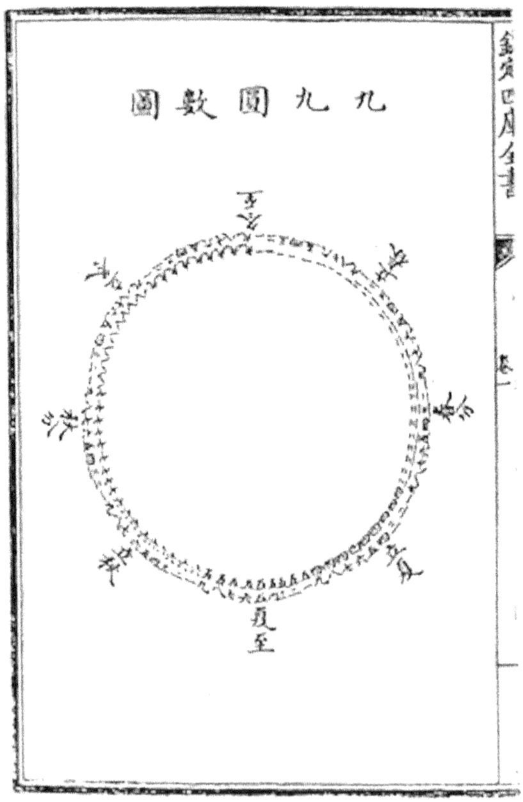

Figure 1.3: Nine by Nine Circular Number Graph

At 45 degree intervals—falling at the 8 doubled figures, 1.1, 2.2, etc.[43]—he indicates solar calendar analogs. These are the 2 solstices, winter (*dongzhi* 冬至) and summer (*xiazhi* 夏至) at 0 and 180 degrees; the 2 equinoxes, spring (*chunfen* 春分) and autumn (*qiufen* 秋分) at 90 and 270 degrees; and, interspersed at 45 degree intervals, the 4 days that mark the advent of each season: advent of spring (*lichun* 立春), summer (*lixia* 立夏), autumn (*liqiu* 立秋), and winter (*lidong* 立冬). Numbers 1.1 and 9.9 share the winter solstice, the former seen as taking up at the point where the latter ends.[44]

In a square version that follows the circular chart in Cai's book, "Nine by Nine Numbers in Rows Chart (*jiujiu hangshutu*, 九九行數圖)," (Figure 1.4), he places 1.1 in the upper right-hand corner:[45]

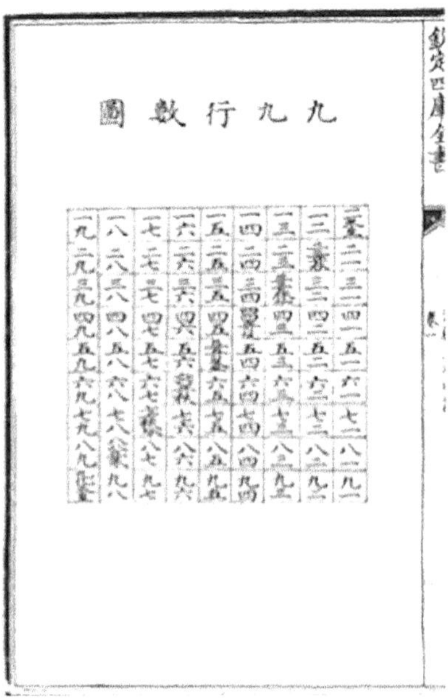
九九行數圖

1 9	1 8	1 7	1 6	1 5	1 4	1 3	1 2	1 1
2 9	2 8	2 7	2 6	2 5	2 4	2 3	2 2	2 1
3 9	3 8	3 7	3 6	3 5	3 4	3 3	3 2	3 1
4 9	4 8	4 7	4 6	4 5	4 4	4 3	4 2	4 1
5 9	5 8	5 7	5 6	5 5	5 4	5 3	5 2	5 1
6 9	6 8	6 7	6 6	6 5	6 4	6 3	6 2	6 1
7 9	7 8	7 7	7 6	7 5	7 4	7 3	7 2	7 1
8 9	8 8	8 7	8 6	8 5	8 4	8 3	8 2	8 1
9 9	9 8	9 7	9 6	9 5	9 4	9 3	9 2	9 1

Figure 1.4: Nine by Nine Numbers in Rows Chart (left)
Conversion to Arabic Numerals (right)

Although this is essentially the same "sequence of numbers from 1 to 81 in nine parallel columns" that Cammann noted as a step in the evolution of the order nine magic square in Yang Hui's book, the presentation exposes two layers of numbers, that are each half of a semi-magic square. Cai took advantage of that doubling to make all his matrices in effect two coordinated matrices, each with its own magic square properties. Note that the eight solar terms here follow the diagonal of doubled terms, 1.1, 2.2, . . . 9.9, and the cyclical flow follows from top to bottom of column 1 back to the top of column 2 and on to its bottom and so forth. Thus, the square is actually a circle whose segments are ranked sequentially in its columns. A vertical 5 axis is marked in the bottom terms of the fifth column and the top terms of the fifth row. A tens complement structure can be observed across the fifth row and column in the top terms of each column and bottom terms of each row and both terms of the diagonals (though the terms in the top left to bottom right diagonal go in opposite directions). In these ways, the most important of the rules Cai applied in constructing his matrices are already present in the base nine natural number sequence as portrayed in preliminary round and square graphics.

D. Structure of the 81 Number Name Matrices

Cai demonstrated his number theory graphically in a suite of 81 matrices, one assigned to each member of the sequence of numbers shown in a matrix titled "Complete Names of the Numbers in the Supreme Pivot Inner Chapters (*Huangji neipianshu zongming* 皇機內篇數總名)."[46] (These names are translated in Table 2.1.) Figure 1.5 shows the matrix for the first Number Name, 1.1 Origin.

The chart is read in the traditional Chinese format of top to bottom, right to left. Note that the top digit of each pair in the Nine by Nine Numbers in Rows Chart is on the right in this configuration and that they are shown in one of the traditional counting rod formats （*suanchou jishufa* 算籌記數法） in place of standard numbers (Table 1.8).[47]

Notation (1) in the upper right-hand corner of Figure 1.5 is the counting rod identifier, "丨 丨," placed above the Number Name, 原 (2), and (3) its family identifier, 一之一, for "the first of family one (i.e., 1.1)". A legend, which begins by repeating the Number Name (4) is followed by the words (5) "the Numbers say (*shuyue* 數曰)."

Origin is the only Number Name that has additional explanatory information after "the Numbers say"; however, in all other respects, its format is the same as all other matrices. Each cell is denominated in counting rod numerals as at arrow (6), here the numbers 9 and 5 (9.5). These numbers run sequentially top to bottom and begin again at the top of the adjacent column to the left from 1.1 in the upper right-hand corner to 1.9 in the lower right-hand corner, 2.1 at the top of the second column to the left, and so on to 9.9 in the lower left-hand corner. This numbering scheme is reproduced in each

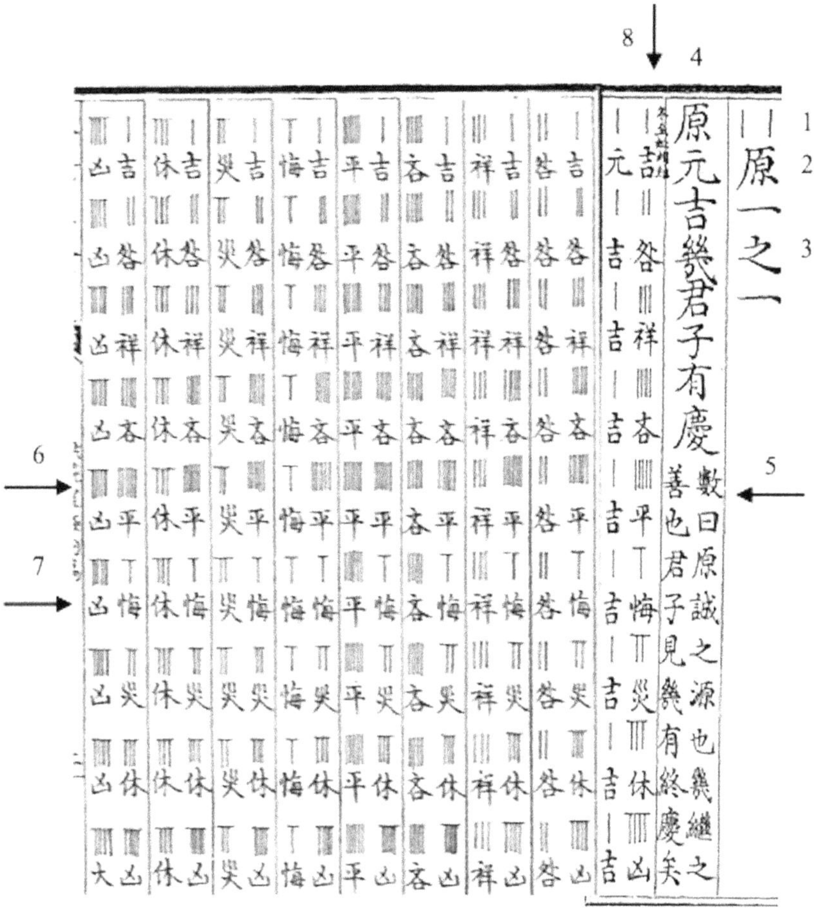

Figure 1.5: Elements of Cai Chen Matrices

𝍠	𝍡	𝍢	𝍣	𝍤	𝍥	𝍦	𝍧	𝍨

Table 1.8: Counting Rod Numerals 1–9

matrix. However, each is different from the others in the distribution of paired terms from Cai's Luck Schedule. Arrow (7) points to one of those pairs *xion-ghui* 凶悔, which translates "Bad Luck.Remorse." These pairs in their unique locations in each matrix are presumably at least in part what Cai intended by prefacing to them the words "the Numbers say." The implication is that he took spatial and mathematical relationships among the pairs and their coordinates on the underlying numerical grid as an input to the diction he chose for the Number Name and legend.

The bipartite numeric associations N.n, where N is the left-hand and n the right-hand term, were applied in a system of divination Cai modeled on the yarrow stalk method associated with the *Yijing*. The procedure, described in Appendix IV, had two separate operations that each yielded one count of one through nine "on the left" and another "on the right."[48] In the first operation, the count gives a "big number (*dashu* 大數), which keys to one of the 81 Number Names in the order shown in Table 2.2 below. Each Number has a legend that is taken to further define its relative degree of good or bad luck. For example, 1.1 reads "Origin. Prodigious Good Luck. All is incipient. Noble-minded person—congratulations."[49]

The second step of the divination procedure again yields two numbers one through nine—the "little number (*xiaoshu* 小數)." These are applied to deriving two terms from Cai's Luck Schedule. The 81 possible combinations of these terms are distributed in pairs in a unique matrix for each of the 81 Number Names, though not all combinations are used in every matrix. The resulting matrices are individualized spectrums of Luck Schedule terms, and it appears that Cai viewed them as meaningfully related to the Number Names and their legends. After each legend, Cai's text has the words "the numbers say (*shu yue*)." For the first Number Name, Origin, he follows that phrase with several definitions and explanations (translated in Chapter 2, 1.1 Origin). Yao Yong, a scholar of the next generation, says Cai died before completing the rest of this self-commentary.[50] But the words "the Numbers say" still appear immediately before each of the 80 remaining matrices with no further comment, implying that Cai perhaps saw patterns reflected in the numerical structure of the matrix that triggered the verbal associations in the legend.

That the changing overlay of luck terms in each matrix traces the seasonal round is emphasized throughout the series by references in small print to the agricultural calendar (*nongli* 農曆) via an almanac in the "Lunar Seasons (*Yueling* 月令)" chapter of *Record of Rites* (*Liji* 禮記) (as in Figure 1.5). These offer iconic imagery for "72 periods (*qishierhou* 七十二候)," which include the familiar 24 "solar terms (*ershisi jieqi* 二十四節氣)". Cai placed these in the same location in the matrices of each N term family, those cells whose right- and left-hand numbers in the underlying grid are the same—starting at cell 1.1 in the matrix for Number Name 1.1, then 2.2 in matrix 1.2, through cell 8.8 in matrix 1.8, and returning to cell 1.1 in matrix 2.1 to resume the cyclical pattern through cell 8.8 in matrix 9.8.[51] Cai gets around the mismatch between the number of periods, 72, and his 81 matrices by not using a seasonal image in any N.9 cell, although, as noted above, he elsewhere asserts that Matrices 9.9 and 1.1 share the Winter Solstice. The rest fall along the 1.1–9.9 diagonal in successive matrices as do Cai's 8 solar terms on the "Nine by Nine Numbers in Rows Chart" in Figure 1.4. As a corollary, they are always in the same row as the AA cell, discussed below (all matrices are shown in Chapter 2).

The underlying numerical counting rod grid seen in Figure 1.5 is the same for all matrices: the numbering runs right to left from " | | ," 1.1, in column 1, row 1 to "𝍨 𝍨,"9.9, in column 9 row 9. However, Cai utilizes a replacement strategy foreshadowed in the Luck Schedule matrix of Table 1.5 to differentiate among the 81 matrices. Thus, each is a separate spectrum of binary terms, though Cai's rules of construction cause them to fall into families based on the left-hand (N) terms of the Number Names, i.e., 1.1–1.9, 2.1–2.9 . . . 9.1–9.9. The rules also create other relationships across the series, as discussed below.

E. Structure of the 81 Matrix Suite

As in parsing the Luck Schedule replacement strategy, understanding the structure of the 81 Number Name matrices is facilitated by conversion to more manageable symbols. In order to differentiate them from the numbering of the Number Names and the rows and columns of the underlying grid, which are given in numerics, the pairs of Luck Schedule terms in Number Name matrices are indicated in the discussion that follows by letters A through I. The spectrum of luck terms for Number Name 1.1, Origin, is:

Table 1.9: 1.1 Origin Matrix[52]

9	8	7	6	5	4	3	2	1	
IA	HA	GA	FA	EA	DA	CA	BA	AA	0.1
IB	HB	GB	FB	EB	DB	CB	BB	AB	0.2
IC	HC	GC	FC	EC	DC	CC	BC	AC	0.3
ID	HD	GD	FD	ED	DD	CD	BD	AD	0.4
IE	HE	GE	FE	EE	DE	CE	BE	AE	0.5
IF	HF	GF	FF	EF	DF	CF	BF	AF	0.6
IG	HG	GG	FG	EG	DG	CG	BG	AG	0.7
IH	HH	GH	FH	EH	DH	CH	BH	AH	0.8
II	HI	GI	FI	EI	DI	CI	BI	AI	0.9

The replacement strategy makes each Number matrix a member of a family related through the left-hand term of their rows. The rows of 1.1–1.9 share the left-hand terms A, B, C, D, E, F, G, H, I, 2.1–2.9 share D, A, B, G, E, C, H, I, F, etc., with the 5.1–5.9 family being an exception in this and other ways (as described below). In 1.1 Origin, right-hand terms follow the order of column 1 in the Luck Schedule, Table 1.5 – i.e., they run in standard alphabetic order, A to I. The right-hand terms follow this same order in all rows. Thereby, the left-hand term remains the same for all cells in a column, and the right-hand term is constant for all cells in a row. This principle holds for all matrices except the fifth (or N.5 matrix) in each family. Note also that the diagonal from 1.1 to 9.9 has the doubled terms in alphabetic order and the diagonal from 9.1 to 1.9 has alphabetic order running forward in the right and reverse in the left term. In all subsequent matrices in the family comprising 1.1–1.9 (except 1.5) left-hand row terms are the same. The sequence of right-hand row terms changes, however, in keeping with columns 2–9 of the Luck Schedule.

Table 1.10: 1.2 Hiding Matrix

9	8	7	6	5	4	3	2	1	
ID	HD	GD	FD	ED	DD	CD	BD	AD	0.1
IA	HA	GA	FA	EA	DA	CA	BA	**AA**	0.2
IB	HB	GB	FB	EB	DB	CB	BB	AB	0.3
IG	HG	GG	FG	EG	DG	CG	BG	AG	0.4
IE	HE	GE	FE	EE	DE	CE	BE	AE	0.5
IC	HC	GC	FC	EC	DC	CC	BC	AC	0.6
IH	HH	GH	FH	EH	DH	CH	BH	AH	0.7
II	HI	GI	FI	EI	DI	CI	BI	AI	0.8
IF	HF	GF	FF	EF	DF	CF	BF	AF	0.9

In matrix 1.2 (Table 1.10), the replacement scheme from the Luck Schedule is invoked in the right-hand terms, replicating the move from column 1 to column 2 in the Luck Schedule. AA and AB along with their entire rows move down one cell to 1.2 and 1.3, AG moves up three cells to 1.4, AE remains in place in 1.5, AC moves down three to 1.6, AH and AI move up one to 1.7 and 1.8, and AF moves down three to 1.9.

This replacement strategy continues for Number Spectrums 1.3, whose right-hand terms in column 1 are the Luck Schedule's column 3, 1.4 (col. 1 = Luck Schedule col. 4), and 1.6 through 1.9. At 1.9 Danger Spectrum, the Figure is the inverse of 1.1. Significantly, these moves place AA in cell 1.2 of matrix 1.2, i.e., the cell with the same numerical value as its matrix. Maintaining this "equivalence rule" throughout the suite of 81 matrices is Cai's primary objective, accommodated by his replacement strategy.

Table 1.11: 1.5 Straightness Matrix

9	8	7	6	5	4	3	2	1	
II	HF	GC	FH	EE	DB	CG	BD	AA	0.1
IH	HI	GF	FG	EE	DC	CD	BA	AB	0.2
IG	HH	GI	FD	EE	DF	CA	BB	AC	0.3
IF	HC	GB	FI	EE	DA	CH	BG	AD	0.4
IA	HA	GA	FA	EA	DA	CA	BA	**AA**	0.5
ID	HG	GH	FA	EE	DI	CB	BC	AF	0.6
IC	HB	GA	FF	EE	DD	CI	BH	AG	0.7
IB	HA	GD	FC	EE	DG	CF	BI	AH	0.8
IA	HD	GG	FB	EE	DH	CC	BF	AI	0.9

The structure of matrix1.5 (Table 1.11) exhibits a secondary tactic that allows the maintenance of Cai's AA equivalence rule in the context of the exceptionality of the number 5 and its alphabetic term, E. 1.5 is derived directly from the full Luck Schedule and the first member of its family, matrix1.1, rather than from the family member that precedes it.

In 1.5 and all N.5 matrices, the right-hand terms in all columns and rows are equivalent to the terms in corresponding cells in the Luck Schedule matrix, Table 1.5 (except in row 5). Left-hand terms are the same as in the rows in same-family matrix N.1, in this case 1.1 Origin's alphabetic order A through I. In the first row, the right-hand terms are the first, fourth, seventh, second, fifth, eighth, third, sixth, and ninth, and so on for all rows except for row 5. There, a row with A as the right-hand term replaces the E row in the Luck Schedule matrix. The left-hand terms remain the same as in the other rows. The Es that are thereby eliminated from that row become the right-hand term in column 5, which shows EE in every cell except 5.5, "EA." This complicated maneuver ensures that AA occurs in the cell that has the same numerical value as the matrix, though in the fifth member of a family, there are always two AAs. In this case the other AA is in cell 1.1, and in other N.5 matrices it occurs on the diagonal from 1.1 to 9.9. The integrity of the semi-magic square across the horizontal and vertical axes is maintained although the N.5 row is mediated by the A term rather than having E as its right-hand term.

Table 1.12: 2.1 Achievement Matrix

9	8	7	6	5	4	3	2	1	
FA	IA	HA	CA	EA	GA	BA	**AA**	DA	0.1
FB	IB	HB	CB	EB	GB	BB	AB	DB	0.2
FC	IC	HC	CC	EC	GC	BC	AC	DC	0.3
FD	ID	HD	CD	ED	GD	BD	AD	DD	0.4
FE	IE	HE	CE	EE	GE	BE	AE	DE	0.5
FF	IF	HF	CF	EF	GF	BF	AF	DF	0.6
FG	IG	HG	CG	EG	GG	BG	AG	DG	0.7
FH	IH	HH	CH	EH	GH	BH	AH	DH	0.8
FI	II	HI	CI	EI	GI	BI	AI	DI	0.9

Each succeeding family N.2–4, 6–9 begins with an adaptation to its rows reflecting its corresponding column in the Luck Schedule. Left-hand terms in the 2.1 Achievement Spectrum (Table 1.12) are the sequence in column 2 in the Luck Schedule. In 3.1, they are the same as Luck Schedule column 3, etc. This is another way of saying that the -3, -3, +1 *Luoshu* operation has been imposed upon their rows relative to the corresponding member of the preceding family. Right-hand terms in each column of matrix 2.1 remain equivalent to Luck Schedule column 1; in 2.2, they are equivalent to Luck Schedule column 2, and so on. Left-hand terms are the same in all columns; right-hand terms are the same in all rows.

Table 1.13: 2.5 Constancy Matrix

9	8	7	6	5	4	3	2	1	
FI	IF	HC	CH	EE	GB	BG	AD	DA	0.1
FH	II	HF	CG	EE	GC	BD	AA	DB	0.2
FG	IH	HI	CD	EE	GF	AB	AB	DC	0.3
FF	IC	HB	CI	EE	GA	BH	AG	DD	0.4
FA	IA	HA	CA	EA	GA	BA	**AA**	DA	0.5
FD	IG	HH	CA	EE	GI	BB	AC	DF	0.6
FC	IB	HA	CF	EE	GD	BI	AH	DG	0.7
FB	IA	HD	CC	EE	GG	BF	AI	DH	0.8
FA	ID	HG	CB	EE	GH	BC	AF	DI	0.9

Similarly, the 2.5 Constancy spectrum (Table 1.13) is a successor to 1.5 Straightness by replacement of the first row's left-hand terms with the order from the second column of the Luck Schedule matrix: the right-hand terms throughout are the same as those in the Luck Schedule template with the exception of the fifth row, where E is again replaced by A in all cells but 5.5.

Matrix families 3, 4, and 6 through 9 follow this pattern of replacement with reference to the Luck Schedule. They can be identified in general by the left-hand term of their first cell as in Table 1.14.:

Table 1.14: Matrix Family N, Cell 1.1 Left-Hand Term

N=9	N=8	N=7	N=6	N=4	N=3	N=2	N=1
I	F	C	H	B	G	D	A

Note that Cai's design gives the N.5 matrices 26 A terms as compared to 17 each of the other terms. This imbalance is furthered in the N=5 family and will be discussed in detail in the "Probabilities of Occurrence" and "Trends of Luck" sections below.

Table 1.15: 5.1 Abundance Matrix

9	8	7	6	5	4	3	2	1	
IA	HA	GA	FA	**AA**	DA	CA	BA	AA	0.1
IB	HB	GB	FB	AB	DB	CB	BB	AB	0.2
IC	HC	GC	FC	AC	DC	CC	BC	AC	0.3
ID	HD	GD	FD	AD	DD	CD	BD	AD	0.4
IE	HE	GE	FE	AE	DE	CE	BE	AE	0.5
IF	HF	GF	FF	AF	DF	CF	BF	AF	0.6
IG	HG	GG	FG	AG	DG	CG	BG	AG	0.7
IH	HH	GH	FH	AH	DH	CH	BH	AH	0.8
II	HI	GI	FI	AI	DI	CI	BI	AI	0.9

Matrices 5.1–5.4 and 5.6–5.9 are derived from matrices 1.1–4.4 and 6.6–9.9, respectively. Thus, matrix 5.1 (Table 1.15) is the same as matrix 1.1 except for column 5.

Column 5 is a repeat of column 1—the column whose left-hand term is A. This places an AA in cell 5.1 in keeping with the progression of AA through the 81 matrices in the cell equal to the current matrix's number. Similarly, matrix 5.2 is the same as matrix 2.2 with the column whose left-hand term is A (in this case, col. 2) repeated in the fifth column. This repetition in column 5 of the column with the left-hand A term column is consistent for members 5.1–5.4 and 5.6–5.9 of the N = 5 family. These eight matrices, therefore, each have nine more As and nine fewer Es than the remaining terms.

Table 1.16: 5.5 Center Matrix

9	8	7	6	5	4	3	2	1	
EI	EF	EC	EH	AE	EB	EG	ED	EA	0.1
EH	EI	EF	EG	AE	EC	ED	EA	EB	0.2
EG	EH	EI	ED	AE	EF	EA	EB	EC	0.3
EF	EC	EB	EI	AE	EA	EH	EG	ED	0.4
EA	EA	EA	EA	**AA**	EA	EA	EA	EA	0.5
ED	EG	EH	EA	AE	EI	EB	EC	EF	0.6
EC	EB	EA	EF	AE	ED	EI	EH	EG	0.7
EB	EA	ED	EC	AE	EG	EF	EI	EH	0.8
EA	ED	EG	EB	AE	EH	EC	EF	EI	0.9

Finally, 5.5, the central matrix of the series (Table 1.16), is unique in having an E term present in every cell except its 5.5, which is AA. Right-hand terms are those of the Luck Schedule matrix except in cell 5.5.

In column 5, the Es are the right-hand terms; in all other columns, E is on the left. Except for the AA in 5.5, column 5 and row 5 are all AE and EA, respectively. Row 5, however, is like the N.5 matrices in having the additional A right-hand term. This proliferation of Es—79 in all—compensates for some of the missing Es in the other N = 5 family members. There are also 27 A terms, adding further to the disproportionate number of that term in other members of the N = 5 family and the N.5 members of other sets.

F. Summary of Rules of Construction for the 81 Matrix Suite

Cai's general objectives in constructing his 81 matrices appear to have been (1) to produce a unique matrix for each of his 81 Number Names; (2) to link the AA pairing to the succession of matrices (AA equivalence); (3) to maintain the centrality of the E term in the median axes (central five rule); (4) to enforce the tens complement rule; and (5) to enforce the odd-even complement rule. The first two objectives require serial adjustments that work around the fixed discipline of the third by applying a numerical replacement strategy apparently drawn from the *Luoshu* arrangement. Perforce, these requirements create a series that as a whole reflects the structure of each individual matrix—the series pivots around matrix 5.5, which effectively acts as an equals sign for the series, as each matrix N pivots around its own equals sign, cell N.5.

In achieving his objectives, Cai followed construction rules that can be deduced in the first instance from the column replacement strategy of the Luck Schedule matrix. The following is a summary of the rules as they are successively evident in (A) the Luck Schedule, (B) the 9 x 9 Number Name matrix that constitutes the underlying grid for all matrices, and (C) the 81 Number Name Luck Spectrum series.

A. Luck Schedule Rules:
   1. Treat Luck Spectrum verbal terms as the numeric equivalents one through nine specified by Cai Chen.
   2. Read columns from right 1.n–9.n; read rows from top N.1 to bottom N.9.
   3. In resulting 9 x 9 matrix, cells N.n occur at intersections of columns and rows.

4. Five occurs in the central row and column forming horizontal and vertical axes (central 5 rule).
5. Other numbers are paired with their tens complements across and equidistant from the two 5 axes in the same rows and columns (tens complement rule).
6. Odd numbers appear in odd-numbered cells in odd-numbered rows and columns and in even numbered cells in even numbered rows and columns; vice-versa for even numbers (odd–even complement rule).
7. Numbers other than five change in column-to-column replacement formula governed by a clockwise (+) and counterclockwise (-) -3, -3, +1 pattern taken from the *Luoshu* graphic (Table 1.1), where number two is three places counterclockwise on the graphic's perimeter from number one, number three is three places counterclockwise from number two, and four is one place clockwise from three.
   (a) Translate circular pattern to column cells as upward or downward replacement;
   (b) In columns 1–3, replacement pattern begins where predecessor ended: (one in col. 1), four (col. 2), seven (col. 3; see Table 1.6);
   (c) Begin column 4. with two instead of eight to maintain *Luoshu* pattern.

B. Incorporating A) and evolving therefrom are the rules for the 9 by 9 Number Name matrix (underlying template):
   1. Employ bipartite base nine numbering to create count 1 (1.1) –81 (9.9).
   2. Populate each cell with a sequential number N.n. (originally in counting rod format, see Figure 1.5).
   3. Populate each cell with an arbitrary Number Name (Tables 2.1 and 2.2).

C. Incorporating B) and evolving therefrom are the rules for the 81 Number Name Luck Spectrum matrices:
   1. Construct atop underlying template used in A) and B) above a series of 81 matrices 1.1–9.9 and label each with equivalent Number Names from B.3).
   2. Treat left-hand term as column designation and right-hand term as row designation.
   3. Make matrix 1.1 the original format for the series of 81 matrices:

   (a) Make its right-hand term in column 1 run A through I top to bottom by row.
   (b) Make the right-hand term in every row the same.
   (c) Make its left-hand terms run A through I left to right.
   (d) Make the left-hand term in every column the same.
4. Evolve matrix 1.2–1.4 and 1.6–1.9 successively from its immediate predecessor:
   (a) Replace right-hand term in column 1 with next term as in rule A.7).
   (b) Apply the latter rule to all N.2–N.4, N.6–N.9 matrices.
8. Evolve matrix 2.1 from 1.1 and 1.2:
   (a) Make the right-hand term of 2.1 column 1 equivalent to the right-hand terms in matrix 1.1.
   (b) Make the left-hand term of 2.1, row 1 equivalent to the right-hand term in 1.2, column 1.
   (c) In this fashion, evolve all Matrices N.1 from 1.1 and 1.n.
9. Create matrix 1.5:
   (a) Make the right-hand column terms equivalent to all terms in the Luck Schedule matrix (A):
      (i) Except in row 5, where place an additional 1A term as right-hand term (making cell 1.5 = AA).
   (b) Make the left-hand column terms equivalent to those in matrix 1.1.
   (c) Create remaining Matrices N.5 by enforcing rule 9a) and exception a.i) and using left-hand terms in all columns from its family head N.1.
10. Evolve Matrices 5.1–5.4 and 5.6–5.9, respectively, from Matrices 1.1–4.4 and 6.6–9.9 by replacing the left-hand term of their column 5 (E) with an A (effectively doubling column N.1);
   (a) For example, matrix 5.1 is matrix 1.1 with As replacing Es as left term in column 5, duplicating column 1; 5.2 is 2.2 with As replacing Es as left term in column 5, duplicating its column 2, etc.).
11. Evolve matrix 5.5 from right-hand terms equivalent to the Luck Schedule (and the right-hand terms of all Matrices N.5):
   (a) Make all left-hand terms = E.
   (b) Except in column 5, where make left-hand term = A.

G. Architecture and Meaning in the 81 Matrix Suite

The entire series of 81 double term matrices displays the underlying structure of each component individual matrix and of the Luck Schedule single term matrix. In the latter, each column is the invert of the column equidistant from the vertical axis, column 5. Likewise, in each family of double term matrices, all terms invert from N.n to N.(10-n) across N.5. For example, matrix 1.9 equals 1.1 with its rows inverted; 2.7 equals 2.3 with rows inverted. In the series of 81 matrices, the double terms in matrices equidistant from matrix 5.5 are reversed and inverted—i.e., (10-N).(10-n) is the reverse invert of N.n. Thus, matrix 9.9 is the reverse invert of 1.1, 2.7 is the reverse invert of 8.3, and so on through the series. This makes each cell of the first 40 of the matrices opposite to the corresponding cell of the matrix equidistant from matrix 5.5 in the second 40. In this way, the tens complement rule informs the entire series and all its cells.

Matrix 5.5 is the odd matrix between the 40 on either side. It has no counterpart; instead, it is, like the vertical and horizontal 5 axes in the Luck Schedule, the crossing point that mediates the complementariness of the four families of nine matrices on either side. Matrix 5.5, having E terms as its left-hand terms in all but column 5 and the Luck Schedule for its right-hand terms, gives the clearest visualization of the compound nature of the 81 matrices. Taken separately, the 162 terms of any matrix form two matrices—individually constituting half a magic square. This duality figures into the successive changes among matrices in their families, as left-hand terms remain stable while right-hand terms undergo the *Luoshu*-based displacements that give each matrix its individuality. From that perspective, with one modification in its N.5 member, families 1–4 and 6–9 share one matrix of left-hand terms and display 9 individualized matrices in right-hand terms.

By sharing its right-hand term with all other N.5 matrices, matrix 5.5 links to the families 1–4 and 6–9 each through their central family member. Meanwhile, the left-hand terms of the N.5 matrices is shared with all other members of its immediate family, which is thereby united in the most granular of their pairs of terms. Thus, each N.5 serves as the integrator of its native family, mediating structurally the four family members on either side. Finally, each N = 5 family matrix is identical to its corresponding N family (5.1 to 1.1, 5.2 to 2.2, etc.) with the exception that in the left-hand terms of N = 5 family matrix fifth columns, an A replaces the E in the correspondent. This nesting of interrelationships is illustrated in Figure 1.6.

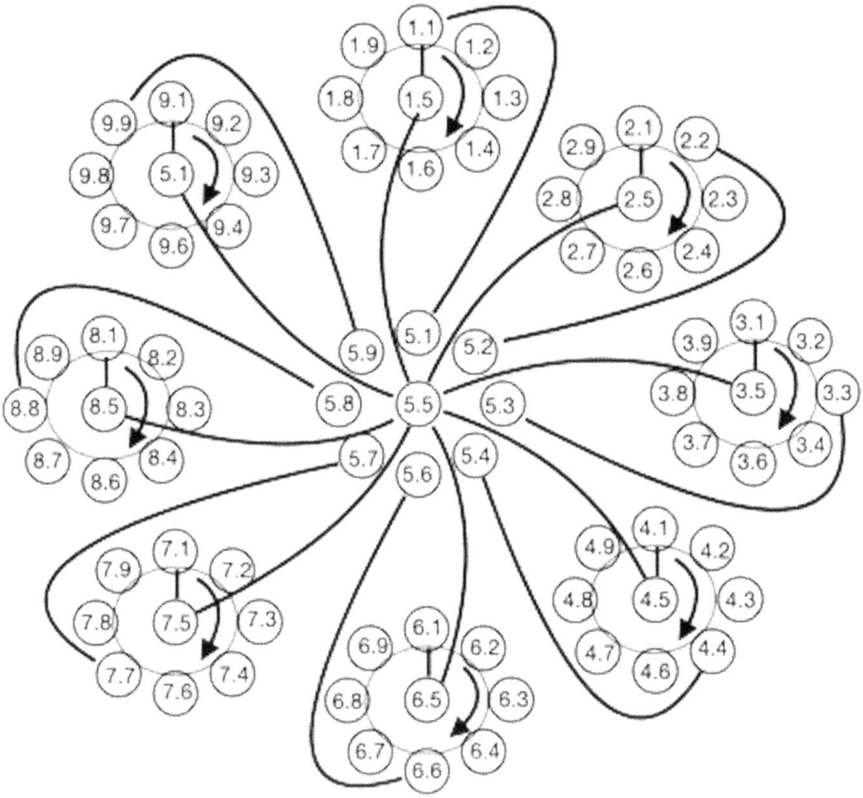

Figure 1.6: Interrelationships of the 81 Matrices

H. Probabilities of Occurrence Among the Luck Schedule Terms

Cell contents in the 81 matrices are produced by the second step of Cai Chen's divination method (see Appendix IV), where a pair of terms from the Luck Schedule is determined by stalk sorting as the "small number (*xiaoshu*)." In Cai's system, the probability of drawing any given term were equal. However, his construction rules necessitate that there are significantly more A terms than any of the other seven. The A term represents Good Luck, and the difference makes the probability of the occurrence of at least one "Good Luck" larger than for any other term. However, the net positive effect that this disparity might suggest for divination is mitigated by other features of both the construction strategy and Cai's divination methodology. Table 1.17 shows the overall distribution of terms in the 81 matrices.

Table 1.17: Distribution of Terms in 81 Number Name Spectrums

Term	Family 1–4,6–9 (no N.5)	Family 1–4,6–9 (no N.5) total	N.5s in Family 1–4,6–9	Family 1–4,6–9 N.5 total	5 Family (no 5.5)	5 Family (no 5.5) total	5.5	Total	% of total	Term-avg. of other terms	5 Family +N.5s	% of term in (5 Family + N.5s)
A	72 x 2 x 8	1152	26 x 8	208	28 x 8	216	27	1603	12.216 %	163	451	16.38 %
E	72 x 2 x 8	1152	17 x 8	136	9 x 8	72	79	1439	10.966 %	-1	287	10.42 %
B	72 x 2 x 8	1152	17 x 8	136	18 x 8	144	8	1440	10.974 %		288	10.46 %
C	72 x 2 x 8	1152	17 x 8	136	18 x 8	144	8	1440	10.974 %		288	10.46 %
D	72 x 2 x 8	1152	17 x 8	136	18 x 8	144	8	1440	10.974 %		288	10.46 %
F	72 x 2 x 8	1152	17 x 8	136	18 x 8	144	8	1440	10.974 %		288	10.46 %
G	72 x 2 x 8	1152	17 x 8	136	18 x 8	144	8	1440	10.974 %		288	10.46 %
H	72 x 2 x 8	1152	17 x 8	136	18 x 8	144	8	1440	10.974 %		288	10.46 %
I	72 x 2 x 8	1152	17 x 8	136	18 x 8	144	8	1440	10.974 %		288	10.46 %
sum		10368		1296		1296	162	13122	79.012 %		2754	20.99 %

The special gestures among the N.5 and 5.n matrices to accommodate the AA equivalence rule result in 9 more A terms and 9 fewer E terms than the remaining terms in each of 5.1–5.4 and 5.6–5.9. Meanwhile, 5.5 has an E term present in every cell except its own cell 5.5, which is AA. This profusion of E terms—79 in all—compensates for some of the missing E terms in the other N = 5 family members. However, there are also 27 A terms in matrix 5.5, adding further to the already disproportionate number of that term in other members of the N = 5 family and the N.5 members of other sets. In all, there are 163 more A terms than any other and one less E. This discrepancy in the relative number of A terms occurs entirely in the N.5 and 5.n matrices, which amount to just over 20 percent of the 81 matrices; so 80 percent of the time, the probability of drawing any given term in divination is equal. But in the N.5 and 5.n matrices the probability of an A appearing is 16.4 percent compared to 10.48 percent for other terms (10.42 percent for E). Overall, an A term would occur 12.2 percent of the time, and other terms close to 11 percent.

However, the luck terms occur in pairs that were apparently taken as a forecast trend in relative good or bad fortune. That is, the first term may have represented the present and the second some point in the future, though as with other aspects of the divination procedure, Cai's intention is not explicit.[53] The pairing AB, for example, would under this understanding imply a trend from Good Luck to Error and to thereby modify with a negative connotation the interpretation of the Number Name and legend drawn in the first operation of the procedure. The reverse pair, BA, would add a positive connotation to the Number Name and legend. Thus, if one's Number Name were 1.1 Origin, the legend, "Origin. Prodigious Good Luck. All is incipient. Noble-minded person—congratulations" is clearly a positive reading. A trend pairing in the second operation of BA—Error-Good Luck—would seem a confirmation of the legend. But a pairing of AB—Good Luck-Error—would seem to imply rather that the positive state would deteriorate. As with any form of divination, the interpretation would presumably depend on a subjective appraisal of the parameters of the situation regarding which the process has been invoked. Cai did not elaborate as to any apparent contradiction between the results of the two operations, but in the *Yijing* tradition on which his work is grounded, it is not uncommon to find a negative line reading within an overall positive hexagram.[54]

Cai applied his understanding of Good Luck as *yang*, odd, positive fortune to the first, third, fifth, and seventh terms of his Luck Schedule. In this, he discounted five from the numbers six through nine so that six would be reduced to five, eight to seven, keeping positive fortune in the odd, *yang* domain. *Yin*—even numbered and negative fortune—associations fall in the second, fourth, sixth, and eighth positions (again discounting five from seven and nine) and partake of negativity. 1A is "Good Luck," 3C "Blessing," and, in spite of 5 intervening, 6F "Regret"[55] and 8H "Felicity" are positive terms. Similarly, numbers 2B "Error," 4D "Remorse," 7G "Disaster," and 9I "Bad Luck" are all treated as even-numbered negative terms as shown in Table 1.18.

Table 1.18: Positive and Negative Qualities of Luck Schedule Terms

	Neutral			(No 5E)	Positive		(No 5E)	Negative
			A	1	Good Luck			
						B	2	Error
			C	3	Blessing			
						D	4	Remorse
E	Balance	5						
			F	-5	Regret			
						G	-6	Disaster
			H	-7	Felicity			
						I	-8	Bad Luck

In the sequence of eight luck terms, the Good and Bad Luck subsequences flow in opposite directions numerically. Good Luck grows from 1 to (7), while Bad Luck diminishes from (8) to 2. These choices reflect Cai's belief that *yang* (odd, Good Luck) advances forward, *yin* (even, Bad Luck) is retrograde.[56] However, the nine terms are an idiosyncratic set used by no one before Cai, and their relative "goodness" or "badness" is not immediately apparent. The presence of "Regret" among the positive terms indicates Cai intended some gradation. And if the sense of paired terms is to indicate a trend of luck, relative strength of luck would seem to matter with respect to interpreting a divination outcome.

However it might have been intended, this pairing mechanism ameliorates to some extent the imbalance that the preponderance of possible A terms might otherwise have on the "fairness" of divination, in the sense of impartiality (as in the notion of "flipping a fair coin"). A "fair" divination system in this sense would offer equal probability of a positive or negative forecast, and Cai Chen's numerically grounded approach—particularly in its second operation to derive a pair of luck terms—offers a test case of his system's overall "fairness." As Table 1.19 shows, the EE and AA pairings have a nontrivial statistical advantage in the distribution of luck types and the 81 possible pairings in the Name Spectrums.

Table 1.19: Frequency of Luck Schedule Terms

Term	Type #	Occurence	# x occ.	# Trend	Total	% Total
(No Trend:)						
AA	1	97	97	1 No		1.48 %
EE	1	128	128	1 No		1.95 %
BB, CC, etc.	7	80	560	7 No		1.22 %
No Trend total occurrences				9 No	785	
No Trend pct. overall total					11.96 %	
(Trend Positive + /Negative -:)						
BA. . .IA, AB. . .IA	14	88	1232	7+/7-		1.34 %
EA	1	88	88	+		1.34 %
AE	1	88	88	-		1.34 %
EB. . .EI, BE. . .IE	14	72	1008	7+/7-		1.10 %
All Other	42	80	3360	21+/21-		1.22 %
+/- total occurrences				36+/36-	5776	
+/- pct. overall total					88.04 %	
overall total	81		6561			

If the 81 possible pairs appeared with equal frequency in the 81 matrices, each pair would appear in $81/81^2 = 1.23$ percent of the cells. Instead, the pair EE appears in 1.95 percent and AA in 1.48 percent of locations. The other doubled terms appear in 1.22 percent of cells—slightly less than the "fair" expectation. Thus, it is most likely that a divination draw would result in a nontrending double Balance (EE) forecast and second most likely that the result would be double Good Luck (AA). Otherwise, there is 1.22 percent likelihood of drawing any other of the doubled terms, all of which forecast no change in the level of luck, though three are on the positive side of the spectrum and four on the negative side. Of the remaining terms the probability of drawing a positive or negative trending pair is equal even though there is a better chance of drawing one of the 16 pairs with at least one A term (including AE and EA at 1.34 percent) and a lesser chance of one of the 14 pairs with one E term (excluding AE and EA at 1.1 percent) than one of the 42 remaining pairs at 1.22 percent. Still, the positive and negative natures of these latter outcomes are equally probable; so to that extent they are "fair" possibilities.

I. Bias Toward Positive Outcomes in 12th-Century Divination Systems

Cai does not address the philosophical implications of this bias in his system. He appears, however, to have been willing to sacrifice "fairness" of the draw for other theoretical considerations—i.e., maintaining the unchanging centrality of the E term and ensuring AA term equivalence in matrix and cell number. This choice resonates with one made by his master, Zhu Xi, who tolerated a 3:1 bias toward "firm" (*yang* or "nine") lines over "weak" (*yin* or "six") lines in his interpretation of the implied yarrow stalk sorting divination procedure in the *Yijing Xici* commentary (see Appendix IV).[57] This *yang:yin* disparity amplifies an inherent proclivity toward Good Luck in the *Yijing's* verbal readings. The ambiguity of the *Yijing*'s language makes it difficult to specify the degree of "luck" in much of its wording. In that context, however, several terms shared by Cai's Luck Schedule are used in the *Yijing* to express relative luck in a divinatory context. The term *ji* 吉 Good Luck is the second most frequent character in the *Yijing*'s hexagram and line texts,[58] and a comparison of its usage versus its opposite *xiong* 凶 Bad Luck offers a reasonable proxy for the general population of terms.[59] There are 147 appearances of Good Luck and 58 of Bad Luck in the 64 sets of hexagram and line words—two and a half times as many.

When the increased probability of sorting out a *yang* line over a *yin* line in Zhu Xi's methodology is taken into account, the bias increases dramatically. A total of 62 unbroken *yang* lines contain the word for Good Luck compared to 56 bipartite *yin* lines. The word Bad Luck appears in 21 *yang* lines compared to 31 *yin* lines, netting 41 chances that a *yang* line will have the more positive reading to 25 for *yin*. As Zhu Xi's sorting method yields a *yang* line 3 times more often than a *yin* line, his approach would make it about 5 times as likely to draw a line betokening Good Luck as Bad.

Zhu Xi did not indicate awareness of the compounding effect of this inherent bias in *Yijing* diction on the mechanics of his stalk sorting methodology. However, he knew that there was a 3 to 1 difference in the outcome of

9 versus 6. He cited Cai Yuanding's naturalistic explanation to rationalize the discrepancy, saying "Of the four seasons, spring, summer, and fall give birth to things and winter does not give birth to things . . . Thus, *yang* is 3 and *yin* is 1."[60] So Zhu Xi did not further pursue a "fair" outcome in stalk sorting, and later scholars' efforts to rectify the imbalance favoring the 9 line were not generally accepted.[61]

Cai Chen responded by inventing a stalk sorting methodology whose direct outcomes were statistically unbiased. However, the structural principles of his matrix construction forced the most probable outcomes of the second operation to be neutral in EE and strongly positive in AA. Also, as with the *Yijing* texts, in Cai's Number legends, "Good Luck" appears 37 times and "Bad Luck" 27. These include 12 legends in which both terms appear, in phrasing that implies, "if X, then Good Luck; if Y, then Bad Luck." The only terms that appear more frequently than "Good Luck" are the negative *bu* 不, 43 times, and the term *li* 利 profit (considered fortunate), 50 times. However, the latter is negated as 不利 19 times, leaving "Good Luck" as the most frequently occurring positive luck term.[62] Thus, like Zhu Xi, Cai ended up with a statistically optimistic system, slightly favoring divination outcomes that any user would likely interpret as favorable.

In summary, Cai Chen claimed that the numbers in his matrices assume verbal form in his Number Names by defining one another in their closed universe cell-by-cell within each matrix and over the entire suite. His rules ensure that numerous layers of interrelationship might be seen to contribute to those definitions—including opposition at a distance over central axes, complementarity in terms of odd and even natures, and changing placement vis-a-vis a standard underlying grid. These dynamics create separate profiles for each matrix, which as a suite trace a circular template that provides a transition to verbal meaning. The annual solar round is a primary source of imagistic content, and other cyclical phenomena offered analogues at shared points on their circumferences. Finally, the strength of the purely numerical interrelationships within the matrix suite seems to be conditioned to a greater or lesser extent by structural choices that affect the probability of their occurrence as the result of a random draw. The choices reflect an optimistic bias that Cai Chen shared with other NeoConfucians of his day—particularly his father and Zhu Xi—and are echoed in his tendency to offer good luck in the wording of his legends.

Notes

1. https://baike.baidu.com/item/书集传.
2. "成吾書者，沈也." *Songshi*, 宋史 列傳193 (ctext.org).
3. Zhu Xi and Joseph Adler, *The Original Meaning of the Yijing*, 11 ff.
4. The *Luoshu* and a similar diagram called the *Hetu*, lit. "River Chart" were incorporated into NeoConfucian thinking in the 11th century by Liu Mu 劉牧 (1011–1064), one of the lines through which the graphics-based number theory of Chen Tuan was transmitted into *Yijing* orthodoxy. Liu Mu originally gave the diagrams the opposite names to the ones they now have. Cai Chen's father, Cai Yuanding, argued for reversing the names and received the sanction of Zhu Xi. See Huang Zongxi, *Yixue xiangshulun*, ch. 1, paragraph 8.
5. "造化之為造化者，幽明屈信而已. . . . 天地也、寒暑也、晝夜也幽明，屈信以成變化者也 What makes Creation Creation is merely dark and light, extension and intension . . .. Heaven and earth, [summer] heat and [winter] cold, day and night all by virtue of darkness and light, extension and intension complete their transformations." Cai, *Hongfan* 1:7a.
6. "動靜無端，後則先矣." *Ibid*., 1:7b.
7. "無形者，理也。That which has no form is Pattern". *Ibid*., 1:10b.
8. "有理，斯有氣。氣著，而理隱。有氣，斯有形。形著，而氣隱。人知形之數，而不知氣之數；人知氣之數，而不知理之數。知理之數，則幾矣。 There is pattern, then there is materiality. Materiality is manifest and Pattern is hidden. There is materiality, then there is a form. The form is manifest and materiality is hidden. People know the numerical [regularity] of the form but may not know the numerical regularity of materiality. Or if people know the numerical [regularity] of materiality, they may not know the numerical [regularity] of Pattern. To know the numerical [regularity] of Pattern is to reach the essence." *Ibid*., 1:7b.
9. See Appendix IV on divination.
10. "體天地之撰者，易之象；紀天地之撰者，範之數。 What embodies the productions of heaven and earth is the *Yi*'s images; what marks the productions of heaven and earth are the numbers of their temporal segments." Cai, *Hongfan*，原序, 1.
11. Described in *ibid*., 4:38b–41b. See Schulz and Thomas J. Cunningham, "Seasonal Structure," 290–291.
12. Nylan, *The Canon of Supreme Mystery*, 15, compares the *Taixuan*'s division of the solar cycle into 81 segments and the *Guaqitu*'s use of the 64 hexagrams.
13. 其法以律起曆，曰：律容一龠，積八十一寸，則一日之分也。*Hanshu*, *Lülizhi* 律歷志, paragraph 20.
14. Cai doesn't refer to them, but his master, Zhu Xi, was interested in both. The latter made numerous references to the *Taixuan*, for example, in *Zhuziyulei*, *Zhouyi* 周易 chapter, *Fuxiguahua xiantiantu* 伏羲卦畫先天圖 section, paragraph 23, and to *guaqi* in *loc. cit*., paragraph 29. The *Sikuquanshu* editors considered Cai's work derivative of both. Cai Chen, *Hongfan*, *Sikuquanshu tiyao*, 3b.
15. 洞極有書，潛虛有圖；非無作也。 *Cai Chen*, *Hongfan Yuanxu*, 1b. Regarding the *Dongji zhenjing* see Huang Zongxi, 4:39b–41b. Regarding the *Qianxu* see *ibid*., 4:32b–38b.
16. 皇极经世 in *ibid*., ch. 5。
17. In *Caishijiuru shu*, 2:23a ff.

18. "蔡氏父子的贡献在于将象数学彻底地从筮法转变为世界图式的建构，并彰"数"在天地万物联系间的作用。" Kang Yu, "*Lun songyuan xiangshusichao*," 94.
19. "先君子曰：洛書者，數之原也." Cai Chen, *Hongfan*, 自序 2a.
20. Cammann, "Evolution of Magic Squares," 118.
21. "漢志曰：禹治洪水，錫洛書法而陳之；洪範是也." Cai Chen, *Shu jizhuan*, 4:19b.
22. "考洛書之名，見於易不見於書。 We observe that the name '*Luoshu*' is seen in the *Yi[jing]* and it is not seen in the *Shu[jing]*. 洪範之文以明理非以明數其事絕不相謀 The language of the [Shujing's chapter] *Hongfan* elucidates principles; it is not about elucidating numbers. The [two] matters are decidedly not compatible." In Cai Chen, *Hongfan*, *Sikuquanshu tiyao*, 2a
23. "九者，九疇之數也。" Cai Chen, *Hongfan*, 原序, 1a.
24. Cammann, "Evolution of Magic Squares," 123.
25. after Cammann, "Evolution of Magic Squares," 117.
26. Cammann, "Evolution of Magic Squares," 121.
27. Cai's divination system is sketched in Cai Chen, *Hongfan*, 2:20a–21b, and described in Appendix IV.
28. Cheng's and other "Hongfan School" offerings are discussed in Appendix III.
29. Cai Chen, *Hongfan*, 自序.
30. *Yijing*, *Xici xia*, paragraph 4. This is taken to mean that a trigram with an odd number of undivided lines (*qian* with three, *zhen*, e.g., with 1) is "*yang*" by nature, while those with none (*kun*) or two (e.g., *xun*) is "*yin*."
31. Cai Chen, *Hongfan* , 2:20a. This passage refers to Cai's divination process described in Appendix IV.
32. *Ibid*., 2:1b.
33. "非一，則不能成兩；非兩，則不能致一。兩者可知，而一者難知也。兩者可見，而一者難見也。可知可見者，體乎？難知難見者，微乎？" *ibid*., 1:12a.
34. "數者始於一，象者成於二。一者奇，二者偶也。奇者，數之所以行。偶者，象之所以立。" *ibid*., 原序, 1a.
35. "數始於一，參於三，究於九，成於八十一，備於六千五百六十一." *ibid*., 原序1a.
36. In the *Hongfan* chapter of *Shujing*, *xiu* 休 is used in a string of *xiuzheng* 休徵 "favorable verifications," in James Legge's translation, for example, "曰肅、時雨若 namely, of gravity, which is emblemed by seasonable rain." In *Yijing* H(exagram) 24 *Fu* 復 6.2：休復，吉, defined by Cheng Yi and Zhu Xi as "休美—with a lovely, pleasing aspect."
37. "一吉，而九凶；三祥，而七災；八休，而二咎；四吝，而六悔；八數周流，推類而求五中." Cai Chen, *Hongfan* , 2.22a.
38. This graphic is first seen in Han Bangqi 韓邦奇 *Hongfan tujie* 洪範圖解, 30b. The reproduction here is from Huang Zongxi 4:48a–b. It does not appear in the *Sikuquanshu* edition of Cai's book.
39. William Symes Andrews, *Magic Squares and Cubes*, 4. Sesiano discusses the possible Indian origin of the kinght's move in *Magic Squares in the Tenth Century*, 6–7. Eric W. Weisstein gives other examples at "Mathematics Notebook, Magic Squares."
40. in 黄宗羲 4:47b. As noted above, the 9 x 9 matrix of the Luck Schedule is also titled with reference to "eight numbers." Various similar progressions through the *Luoshu* numbers

are also common to 風水 Fengshui from an indeterminate date. See Zhuan Hongguang, *Fengshui luopan quanjie*, 121.

41. Cai Chen, *Hongfan*, 1.1b.
42. Note that the 81 matrix series are numbered "[matrix] one of [matrix family] one (*yizhiyi* 一之一), [matrix] two of [matrix family] one一之二, etc. In this regard, the N.n format is a functional translation that identifies column (N) and row (n) locations in a given matrix and also applies to individual members of matrix families.
43. "一數之周，一歲之運也。九數之重，八節之分也。 One round of the [series of] numbers is the progression of one year; in the doubled nine numbers are the divisions of the eight seasonals." Cai Chen, *Hongfan*, 2.3b.
44. "一九首尾為一者，一歲首尾於冬至也。盖冬至二，而餘則一也。 That one and nine at the head and tail are one and the same is like the head and tail of a year both falling at the winter solstice. The winter solstice is two but after all is one." *ibid.*, 2.3b.
45. This illustration and the circular one shown above are based on the hexagram arrangements in Shao Yong's *Fuxi liushisigua fangtu* 伏羲六十四卦方圖 and *yuantu* 圆图. See Zhu Xi and Joseph Adler, *Original Meaning*, 27. Shao's graphic, which is an 8 x 8 matrix, places what he considered the first hexagram in numerical order, *qian* 乾—the analog of Cai's 1.1—in the lower right-hand corner, and Cai makes this comparison explicit in a graphic entitled "Nine by Nine Square Number (*jiujiu fangshutu* 九九方数图)," Cai Chen, *Hongfan*, 1.2a. An appendix to Cai's book shows Shao's combined square and circular arrangements of the 64 hexagrams (*ibid.*, 5:20b) along with a matching arrangement of Cai's 81 counting rod numerical configurations (*ibid.*, 5:22a). A note by 姚鏞 Yao Yong (t. 希聲,1191–?), who also wrote a postface, says: "右二圖得於九峯先生之子抗。有以見八卦九章，相為表裏對待流行。闕一，不可殆與天地終始者乎？因附刻於內篇之後。The two charts on the right were obtained from Mr. Nine Peaks's [i.e., Cai Chen] son Kang, whereby one can see that the eight trigrams and nine chapters [i.e., the number sets of the *Shujing Hongfan* chapter] are mutually a display of opposition and flow through Phases. Are they not decidedly one and unable to be at odds with the beginnings and ends of heaven and earth? Thus we append them at the end of this reprinting of the *Inner Chapters*" (*Ibid.*, 5:23a).
46. In his gloss of the term 皇極 in *ibid.*, 3.1a-b.
47. Sima Guang's *Xuqian*, which Cai Chen acknowledges in his preface, *Hongfan*, 原序 1.b, as an antecedent, earlier employed an alternate form of counting rods.
48. Cai Chen, *Hongfan*, 2.20a-b.
49. 1.1: "原。元吉。幾。君子，慶。" *ibid.*, 3:2a.. The legends are translated in Chapter 2.
50. Yao Yong's postface says: "惜也，不壽而數之辭未備。 It is a pity that he was not long-lived and his wording was not completed" (in Cai, *Hongfan* 5:23b), which implies the belief that Cai intended further explanations to follow those words for each Number Name. There are two passages in the that seem to support Yao's contention. The first begins: "原之一一曰：君子見幾，不俟終日。數曰：知至至之，可與幾也 ." The 1.1 [cell] of [1.1 matrix] Origin says, 'the Noble-minded person seeing the incipient doesn't wait till the end of the day.' The Numbers say, 'One whose knowledge reaches the maximal can be conjoint with the incipient.'" *Ibid.*, 2:11a. These lines are, respectively, quotes from the *Xici xia* paragraph 5 and the *Wenyan* commentary for the third line of H1 *Qian*. The first appears to assign a verbal image to the first cell in the first Number Name, and the second to make that image

a function of the cell's numerical position in the grid. The passage goes on to give additional lines for 中之五五"'Center's' 5.5 [cell]" and 終之九九"'End's 9.9 [cell.]" These also have "Numbers say" passages that cite the *Wenyan* commentary. The second passage again begins: "原之一一者繼之善也 'Origin's' 1.1 [cell] is the continuation of the good," which cites *Xici shang*, para. 5, and goes on to say, "原之九九者，逆而凶也。 Origin's 9.9 [cell] is retrograde and Bad Luck." *ibid.*, 2:11a. As these are presented, they imply the author gave verbal expression to particular cells in the Origin matrix. Perhaps he had in mind another, more detailed layer of wording similar to the *Yijing*'s line words. Some of his followers pursued such a direction. Cheng Zongshun's elaboration of Cai's ideas, *Hongfan jianjie*, presents nine additional readings for each Number Name in *Yijing* fashion, apparently Cheng's own inventions.

51. Although in the *Sikuquanshu* edn., the image is written in cell 8.6, presumably in error as it is the only break in the overall pattern of placement. Cai Chen, *Hongfan*, 5:12a.
52. All of Cai Chen's matrices, shown with alphanumeric counterparts to their original Chinese characters, are adapted from Cai Chen, Hongfan, ch. 3-5. The matrices run sequentially and are numbered in pairs of Chinese characters 一 through 九 separated by the possessive 之：一之一，一之二， 。。。九之九. These are rendered herein 1.1, 1.2. . .9.9.
53. The only example given is in the gloss to Cai's spare description: "扲四揲之後，又加四揲以求其小數而决吉凶。如原之一一則為元吉。 After performing four sorts, add four sorts to seek the small number and determine Good or Bad luck. For example, 1.1 under Origin is Prodigious Good luck." This refers to an AA reading, which is intuitively a "good" reading.
54. For example, the hexagram words of *Yijing*, H7 *Shi* read "師：貞，丈人，吉无咎。Army: Sooth, mature man Good Luck, no Error." But its first, third, and fifth lines all portend Bad Luck.
55. In the *Yijing* commentary tradition, *hui* 悔, Regret, is treated as an opportunity to recognize fault and reform. In *Xici shang*, *paragraph* 2："吉凶者，失得之象也。悔吝者，憂虞之象也 Good Luck and Bad Luck are the images of gain and loss. Regret and remorse are the images of concern and worry." Zhu Xi's gloss reads: "蓋吉凶先對，而悔吝居其中間。悔自兇而趨吉，吝自吉而向凶也。 It seems Good Luck and Bad Luck are first set up as opposites, and Regret and Remorse are between them. Regret moves from Bad Luck towards good Luck; Remorse from Good Luck inclines toward Bad Luck." Zhu Xi, *Yuanben Zhouyibenyi*, 7:2b. Zhu Xi also said, "悔陽而吝陰 Regret is *yang*, Remorse is *yin*. in *Zhuzi yulei*, *Yi 2* 易二, paragraph 51.
56. Cai wrote: "陽順而陰逆；陽長而陰消；陽進而陰退。順者，吉而逆者凶耶. . .以奇偶言，則奇陽而偶. *Yang* is forward, *yin* is retrograde; *yang* grows, *yin* wanes; *yang* advances, *yin* retreats. *Yang* advancing is Good Luck, *yin* retrograde is Bad Luck. . .Speaking in terms of odd and even, then odd is *yang* and even is *yin*." Cai Chen,, *Hongfan*, 1.9b. He further defines "forward" and "retrograde" as "自陰而陽也，順；自陽而陰，也逆。 From *yin* to *yang* is forward; from *yang* to *yin* is retrograde." *Ibid.*, 2.9b.
57. Sun Lesong 孫功松 in his blog post "*Guoshe yu guale—Guo Yong he ZhuXi guanyu shifa de bianlun* 過揲與掛扐—郭雍和朱熹關於蓍法的辯論 (https://blog.xuite.net/dejavu8899/blog/112300197-過揲與掛扐—郭雍和朱熹關於蓍法的辯論+%2F孫功松+%5B轉%5D) discusses the debate between Zhu Xi and his younger contemporary Guo Yong (1106–1187)

regarding the what stalks were to be included in the final count. Guo's discrepancy in favor of the *yang* 9 line were even more extreme than Zhu's, arriving at a ratio of a 9 *yang* line to a 6 *yin* = 27:1.

58. The negative *wu* 无 is most frequent at 158 appearances.
59. Other Luck Schedule terms include 咎 Error, which appears 100 times but in 93 of those it is preceded by a negative implying the fortunate avoidance of Error. 悔 Regret occurs 34 times, and 吝 20 times. 災 Disaster occurs 5 times; 休 Felicity occurs twice (once with Cai's seeming connotation, see note 40), 祥 Blessing occurs once, 平 Balance occurs twice but not as a luck term.
60. "四時春夏秋生物而冬不生物. . .則陽三而陰一也." In Hu fangping 胡方平, *Yixue qimeng tongshi* 易學啟蒙通釋 (四庫全書 edn., archive.org) 卷下, 19a-b.
61. Huang Zongxi discusses alternate views, *Yixue xiangshulun*, 2: 38b–39b.
62. Among Luck Schedule terms 咎 appears twice in Cai's legends, but both in the negated form 無咎. 悔 appears 2 times, 平 2times, 吝 2 times, 休 2 times (but in the same Legend); and 祥 and 災 do not appear. Other terms in legends that commentators take as positive in their *Yijing* contexts are 正 20 times; 亨 10 (twice in the phrase 元亨); 貞 1; 孚 3; 用 11 (though 5 times negated as 勿用); 利有攸往 2. Negative terms include 厲 10; 不利有攸往 4.

· 2 ·

IMAGES

Xiang

象

Expanding a line from Cai Chen quoted partially in the introduction, "Number is the timing of patterns; words are the meaning of number; good luck and bad luck are the determination of words; benefit, advancement, following, and regression are judgements with respect to good luck and bad luck."[1] By "timing," Cai alludes to his circular template whose "pattern" is the solar year and whose seasonal round is repeated in the life cycle of each phenomenal being. The cycle can be quantified—"timed"—in various sets of numbers from the binary, as in *yin* and *yang* or good luck and bad luck, to the 365.25 days in a year and in finer distinction to the 6,561 cells in his suite of 81 matrices. To the nodes in that cyclical model, Cai gave Number Names—his effort to capture the "meaning of number." These meanings can be categorized in general as moments of good or bad luck that come into play as the cycle proceeds, and in the first operation of divination, one Number Name is identified as pertinent to the fortune of the time when it is performed. In the second operation, a judgement of relative luckiness within that moment is given in two terms of Cai's Luck Schedule. And, as discussed in Chapter 1, the term "Balance" holds a desirable place between the two categories in Cai's system and points to the disciplined attitude that allows one to best negotiate the vicissitudes of temporal situations.

This chapter will address the verbal images that Cai Chen evolved in response to his sense of the numerical basis of phenomenal existence. The Number Names are Cai's inventions,[2] the first being 1.1 Origin and the last 9.9 End. Recalling that Cai locates key annual solar periods at the doubled numbers of his "Nine by Nine" graphics (Figure 1.3), all the Number Names in between would seem intended to reflect progressively a node or degree of the annual round. At the same time, analogically, they are taken to represent phases in the developmental cyclical between an entity's coming into and passing out of phenomenal existence. The following Tables show Cai's Number Names and their English translations and adopt the convention of labeling the columns N and the rows 0.n relative to the horizontal and vertical 5 axes. Thus, Origin is 1.1, Hiding is 1.2, . . . End is 9.9:

Table 2.1: Number Names in Chinese

9	8	7	6	5	4	3	2	1	
養	實	迅	飾	庶	比	見	成	原	0.1
遇	賔	懼	戾	決	開	獲	冲	潛	0.2
勝	危	除	虛	豫	晉	從	振	守	0.3
囚	堅	弱	昧	升	公	交	祈	信	0.4
壬	革	疾	損	中	益	育	常	直	0.5
固	報	競	用	伏	章	壯	柔	蒙	0.6
移	止	分	郤	過	盈	興	易	閑	0.7
墮	戎	訟	翕	疑	錫	欣	親	須	0.8
終	結	收	遠	寡	靡	舒	華	厲	0.9

Table 2.2: Number Names in English

9	8	7	6	5	4	3	2	1	
Nourishing	Fruits	Quickness	Decoration	Abundance	Nearness	Visibility	Achievement	Origin	0.1
Meeting	Sojourn	Anxiety	Perversity	Decision	Opening	Taking	Peace	Hiding	0.2
Triumph	Peril	Ridding	Emptiness	Contentment	Progress	Following	Arousal	Guarding	0.3
Imprisonment	Firmness	Weakness	Dimness	Ascenscion	Altruism	Mating	Prayer	Extending	0.4
Cunning	Change	Sickness	Decrease	Center	Increase	Nurturing	Constancy	Straightness	0.5
Rigidity	Repayment	Competing	Utility	Subservience	Distinctness	Strength	Softness	Naiveté	0.6
Shifting	Stopping	Dividing	Withdrawal	Excess	Fullness	Uplifting	Tranquility	Repose	0.7
Falling	Weapons	Dispute	Agreement	Doubt	Bestowing	Gladness	Kinship	Waiting	0.8
End	Knotting	Harvest	Distance	Minimizing	Revelry	Soothing	Flowering	Threat	0.9

As discussed in Chapter 1, Cai's formula for developing the series hinged on the progression of the double Good Luck pair AA cell-by-cell through the series, and that cell is shaded in each matrix below. The AA cell would seem to be the center of gravity in each matrix, and its position may provide an insight into how Cai imagined the numerical configuration as contributing to its matrix's meaning.

Another feature of 72 of the 81 matrices is brief almanac notices in tiny writing in one cell. These three to seven character phrases allude to calendric terms and iconic events in the annual round. Marked with an asterisk, they occur in all but the ninth matrix of each family (N.9) and use the classic set of "72 seasonal periods (*qishierhou*)"—six per month of five days each—of the agricultural calendar (*nongli*). The images are drawn from the "Lunar Seasons (*Yueling*)" chapter of *Record of Rites* (*Liji*)[3] and suggest regularly recurring natural phenomena like "earthworms curl up" in matrix 1.1 and "elk shed antlers" in 1.2. These periodic phenomena belong in turn to one of "24 solar terms (*ershisi jieqi*)"—like "Small Cold (*xiaohan* 小寒)" in matrix 1.4 and "Great Cold (*dahan* 大寒)" in 1.7. In bypassing all the N.9 matrices except 9.9, Cai was left with 72 cells for these annotations, spaced 10 cells apart in each matrix and falling on the 1.1–9.9 diagonal in successive members of matrix families.

In the following translations, I have attempted to achieve consistency with Cai's diction, where possible using the same English term (adjusted for part of speech) for any given character and trying to convey in English the minimalistic classical Chinese phrasing. Such a tactic has limitations—for example, in dealing with the various possible definitions of the character *ming* 明, which can mean light or clarity or brilliance, intelligence or understanding. However, in particular with a divination text, it is useful to see the ways in which ambiguities of definition and grammar create interpretive alternatives for the user to contemplate. It is that quality that allows the mind of the user to invest one set of words with myriad personalized connotations.

Cai used divination terms also found in the *Yijing*, and, as Cai was Zhu Xi's student, I translate these in the sense of Zhu's commentary, the *Zhouyi Benyi*, and highlight them and other Luck Schedule terms by capitalization.[4] Several of the Number Names are rendered with reference to a Qing Dynasty compilation called *Writings of the Nine Confucians of the Cai Family* (*Caishijiuru shu*). This book contains two charts not in the *Sikuquanshu* edition that give a synonym for each Number Name and Luck Schedule term. They proved decisive in giving the English equivalents "Error" in the Luck Schedule and

Number Names, such as 2.2 "Peace," 2.7 "Tranquility," 6.7 "Withdrawal," and 9.5 "Cunning." In addition, numerous legends, beginning with 3.9 Soothing, are rhyming poems, and the rhymes are indicated by ^{R} in the Chinese character line.

Each Number Name is laid out as in Cai's format as depicted in Figure 1.5 and accompanied by its matrix rendered into alphabetic form so that the reader can visualize its scope, refer to the corresponding terms in the Luck Schedule, and have a sense of its numerical structure. Significant references to classical texts are noted as are supporting allusions in Cai's discourses that precede the matrices in his book. *Yijing* hexagrams are cited as H and their number in the *Yijing* order. Where no additional number is given, the citation refers to the hexagram words (e.g., *Yijing*, H4, *Meng* for *Meng* hexagram words). Line words are noted as 9.(1–6) for a "9" *yang* single line or 6.(1–6) for a "6" *yin* bipartite line (e.g., *Yijing*, H4 *Meng* 9.5, for *Meng* hexagram's single fifth line).

Finally, I give a "comment" for each legend, rather in the style of traditional line-by-line commentary in a classic Chinese text. Its purpose is to elaborate the allusive compression and telegraphic syntax of the original and to bring out the self-consistency of Cai Chen's thoughts. These remarks also take a cue from Cai's master, Zhu Xi, who said of studying the *Yijing*'s obscure messages: "People of today must read [the *Yijing*] as a divination text—then they will get what it means."[5] In this spirit, the comments suggest how someone who has drawn any given Number Name as a divinatory reading might understand the prognostication.

A. Chart of the 81 Numbers

Bashiyishu tu

八十一數圖

原 一之一

原。元吉。幾。君子慶。

Origin[6] *Yuan* 1.1

Origin. Prodigious Good Luck. All is incipient.[7] Noble-minded person—congratulations.[8]

數曰：原，誠之源也。幾，繼而善也。君子見幾有終，吉也。

The Numbers say: Origin is the source of sincerity. What is incipient continues on to the good. The noble-minded person sees the incipient has an ending; that is his Good Luck.

9	8	7	6	5	4	3	2	1	
IA	HA	GA	FA	EA	DA	CA	BA	AA*	0.1
IB	HB	GB	FB	EB	DB	CB	BB	AB	0.2
IC	HC	GC	FC	EC	DC	CC	BC	AC	0.3
ID	HD	GD	FD	ED	DD	CD	BD	AD	0.4
IE	HE	GE	FE	EE	DE	CE	BE	AE	0.5
IF	HF	GF	FF	EF	DF	CF	BF	AF	0.6
IG	HG	GG	FG	EG	DG	CG	BG	AG	0.7
IH	HH	GH	FH	EH	DH	CH	BH	AH	0.8
II	HI	GI	FI	EI	DI	CI	BI	AI	0.9

*冬至；蚯蚓結。 Winter Solstice;[9] earthworms coil.

Comment: The origin of any phenomenon, the beginning of a life cycle, is epitomized in the winter solstice, the day of least sunlight. From that turning point, nature will renew its eternal round; and at this moment, what lies ahead is growth and flourishing. These are the seasons of prodigious Good Luck. The noble-minded person understands the underlying Pattern, which proceeds inevitably from growth to decay and ending. This knowledge informs one's mental stance, keying correct action to honest analysis—that is the meaning of sincerity. But in this incipient moment, one can act in anticipation of a positive outcome and receiving congratulations for success.

潛 一之二

潛。勿用有攸往。正靜，吉。

Hiding *Qian* 1.2

Hiding. Do not[10] go anywhere. Propriety, stillness—Good Luck.

數曰

The Numbers say:[11]

9	8	7	6	5	4	3	2	1	
ID	HD	GD	FD	ED	DD	CD	BD	AD	0.1
IA	HA	GA	FA	EA	DA	CA	BA*	AA	0.2
IB	HB	GB	FB	EB	DB	CB	BB	AB	0.3
IG	HG	GG	FG	EG	DG	CG	BG	AG	0.4
IE	HE	GE	FE	EE	DE	CE	BE	AE	0.5
IC	HC	GC	FC	EC	DC	CC	BC	AC	0.6
IH	HH	GH	FH	EH	DH	CH	BH	AH	0.7
II	HI	GI	FI	EI	DI	CI	BI	AI	0.8
IF	HF	GF	FF	EF	DF	CF	BF	AF	0.9

*麋角解。 Elk shed antlers

Comment: After the origin of a life cycle, the incipient being is hidden away in embryonic vulnerability. In proper compliance with this condition, one must remain sheltered in place and keep still to ensure Good Luck.

守一之三

守。居正吉。不利有攸往。

Guarding *Shou* 1.3

Guarding. Dwell in propriety—Good Luck. No profit going anywhere.

數曰

The Numbers say:

9	8	7	6	5	4	3	2	1	
IG	HG	GG	FG	EG	DG	CG	BG	AG	0.1
ID	HD	GD	FD	ED	DD	CD	BD	AD	0.2
IA	HA	GA	FA	EA	DA	CA*	BA	AA	0.3
IH	HH	GH	FH	EH	DH	CH	BH	AH	0.4
IE	HE	GE	FE	EE	DE	CE	BE	AE	0.5
IB	HB	GB	FB	EB	DB	CB	BB	AB	0.6
II	HI	GI	FI	EI	DI	CI	BI	AI	0.7
IF	HF	GF	FF	EF	DF	CF	BF	AF	0.8
IC	HC	GC	FC	EC	DC	CC	BC	AC	0.9

*水泉動。 Springs of water move.

Comment: One phase farther along but still embryonic and in the darkness of winter, one must dwell in the proper response to the moment. Being on guard in one's weakness leads to Good Luck. One remains in watchful repose and will gain nothing from venturing out.

信 一之四

信。中實。有孚。祭祀。

Extending *Shen* 1.4

Extending. The center bears fruit. Have faith. Sacrifice, worship.

數曰

The Numbers say:

9	8	7	6	5	4	3	2	1	
IB	HB	GB	FB	EB	DB	CB	BB	AB	0.1
IC	HC	GC	FC	EC	DC	CC	BC	AC	0.2
IF	HF	GF	FF	EF	DF	CF	BF	AF	0.3
IA	HA	GA	FA	EA	DA*	CA	BA	AA	0.4
IE	HE	GE	FE	EE	DE	CE	BE	AE	0.5
II	HI	GI	FI	EI	DI	CI	BI	AI	0.6
ID	HD	GD	FD	ED	DD	CD	BD	AD	0.7
IG	HG	GG	FG	EG	DG	CG	BG	AG	0.8
IH	HH	GH	FH	EH	DH	CH	BH	AH	0.9

*小寒；雁北鄉。 Small Cold;[12] geese lead toward the north.[13]

Comment: The being to be begins to form in the womb, the future fruit takes shape in the seed in the earth. That it will extend and develop is a matter of faith rather than certainty; thus, one expresses one's faith in the rituals that affirm awareness of the natural cycle—sacrifices and acts of worship traditionally observed at this time of mid-winter.

直 一之五

直。有事，勿事。敬之吉。正凶。利見大人。

Straightness *Zhi* 1.5

Straightness. Matters arise; do not engage in them. Pay homage—Good Luck。 Propriety—Bad Luck. Profit in seeing the great person.

數曰

The Numbers say:

9	8	7	6	5	4	3	2	1	
II	HF	GC	FH	EE	DB	CG	BD	AA	0.1
IH	HI	GF	FG	EE	DC	CD	BA	AB	0.2
IG	HH	GI	FD	EE	DF	CA	BB	AC	0.3
IF	HC	GB	FI	EE	DA	CH	BG	AD	0.4
IA	HA	GA	FA	EA*	DA	CA	BA	AA	0.5
ID	HG	GH	FA	EE	DI	CB	BC	AF	0.6
IC	HB	GA	FF	EE	DD	CI	BH	AG	0.7
IB	HA	GD	FC	EE	DG	CF	BI	AH	0.8
IA	HD	GG	FB	EE	DH	CC	BF	AI	0.9

*鵲始巢。 Magpies first nest.

Comment: What is proper to this still early phase of development is to remain secluded. But events from the outside world intrude and make what would otherwise be proper a source of Bad Luck. One must encounter these events in a straight, frank manner. Acknowledge but do not engage in them. One will profit from the advice of a wise and experienced person and should make an effort to seek such counsel.

蒙 一之六

蒙。小事，吉。內明外蒙。迫則凶。利數學。

Naiveté *Meng* 1.6

Naiveté. In small matters, Good Luck. Inside there is light; outside, Naiveté. Forcing brings Bad Luck. Profit in the study of numbers.

數曰

The Numbers say:

9	8	7	6	5	4	3	2	1	
IH	HH	GH	FH	EH	DH	CH	BH	AH	0.1
IG	HG	GG	FG	EG	DG	CG	BG	AG	0.2
ID	HD	GD	FD	ED	DD	CD	BD	AD	0.3
II	HI	GI	FI	EI	DI	CI	BI	AI	0.4
IE	HE	GE	FE	EE	DE	CE	BE	AE	0.5
IA	HA	GA	FA*	EA	DA	CA	BA	AA	0.6
IF	HF	GF	FF	EF	DF	CF	BF	AF	0.7
IC	HC	GC	FC	EC	DC	CC	BC	AC	0.8
IB	HB	GB	FB	EB	DB	CB	BB	AB	0.9

*雉始雊。 The pheasant first crows.

Comment: Dawning consciousness has only slight understanding—a small light inside that from the outside is seen by others as Naiveté. A *naif* can deal with small matters and have Good Luck. But pressing on to manage more complex situations will lead to Bad Luck. One can at this stage become more sophisticated by studying the numerical structure of natural cycles.

閑 一之七

閑。厲。利禦寇。勿越勿逐。

Repose *Bi* 1.7

Repose. Danger. Profit in resisting bandits.[14] Do not overreach; do not chase.

數曰:

The Numbers say:

9	8	7	6	5	4	3	2	1	
IC	HC	GC	FC	EC	DC	CC	BC	AC	0.1
IF	HF	GF	FF	EF	DF	CF	BF	AF	0.2
II	HI	GI	FI	EI	DI	CI	BI	AI	0.3
IB	HB	GB	FB	EB	DB	CB	BB	AB	0.4
IE	HE	GE	FE	EE	DE	CE	BE	AE	0.5
IH	HH	GH	FH	EH	DH	CH	BH	AH	0.6
IA	HA	GA*	FA	EA	DA	CA	BA	AA	0.7
ID	HD	GD	FD	ED	DD	CD	BD	AD	0.8
IG	HG	GG	FG	EG	DG	CG	BG	AG	0.9

*大寒；雞始乳。 Great Cold;[15] hens first brood.

Comment: In the coldest days of winter, one is in repose within. Still, there is danger without—bandits roam the countryside, and one gains by resisting their incursions. But one must remain on the defensive: it will not do to overreach by going in pursuit of the foe.

須一之八

須。有孚；未明。不利攸行。中正有慶。

Waiting *Xu* 1.8

Waiting. Have faith; there is no light. No profit in taking action. Be centered and proper and receive congratulations.[16]

數曰

The Numbers say:

9	8	7	6	5	4	3	2	1	
IF	HF	GF	FF	EF	DF	CF	BF	AF	0.1
II	HI	GI	FI	EI	DI	CI	BI	AI	0.2
IH	HH	GH	FH	EH	DH	CH	BH	AH	0.3
IC	HC	GC	FC	EC	DC	CC	BC	AC	0.4
IE	HE	GE	FE	EE	DE	CE	BE	AE	0.5
IG	HG	GG	FG	EG	DG	CG	BG	AG	0.6
IB	HB	GB	FB	EB	DB	CB	BB	AB	0.7
IA	HA*	GA	FA	EA	DA	CA	BA	AA	0.8
ID	HD	GD	FD	ED	DD	CD	BD	AD	0.9

*征鳥厲疾。 Birds of prey are swift danger.

Comment: One waits in the dark, cold winter. At this time, one must have faith in the natural order's promise of return to light and warmth. But there is no advantage to be gained by taking action; instead, if one holds one's center and exercises propriety, one will receive congratulations.

厲 一之九

厲。征鳥厲疾。無初有終。吉

Danger *Li* 1.9

Danger. Birds of prey are swift danger.[17] There is no beginning; there is an end[18]—Good Luck.

數曰:

The Numbers say:

9	8	7	6	5	4	3	2	1	
II	HI	GI	FI	EI	DI	CI	BI	AI	0.1
IH	HH	GH	FH	EH	DH	CH	BH	AH	0.2
IG	HG	GG	FG	EG	DG	CG	BG	AG	0.3
IF	HF	GF	FF	EF	DF	CF	BF	AF	0.4
IE	HE	GE	FE	EE	DE	CE	BE	AE	0.5
ID	HD	GD	FD	ED	DD	CD	BD	AD	0.6
IC	HC	GC	FC	EC	DC	CC	BC	AC	0.7
IB	HB	GB	FB	EB	DB	CB	BB	AB	0.8
IA	HA	GA	FA	EA	DA	CA	BA	AA	0.9

Comment: Birds of prey mean can mean sudden death to small animals venturing out into the snow to forage. So also does Danger lurk for one who leaves the place where one waits out the winter. But when the raptor is not present, the danger cannot become a reality—the problem by not arising comes to an end. Is this not Good Luck?

成 二之一

成。正惠有終。吉。不利有攸往。勿首事。毀成，凶

Achievement *Cheng* 2.1

Achievement. Propriety and kindness bring to an ending. Good Luck. No profit in going anywhere. Do not take the lead. Ruining achievement—Bad Luck.

數曰:

The Numbers say:

9	8	7	6	5	4	3	2	1	
FA	IA	HA	CA	EA	GA	BA	AA	DA*	0.1
FB	IB	HB	CB	EB	GB	BB	AB	DB	0.2
FC	IC	HC	CC	EC	GC	BC	AC	DC	0.3
FD	ID	HD	CD	ED	GD	BD	AD	DD	0.4
FE	IE	HE	CE	EE	GE	BE	AE	DE	0.5
FF	IF	HF	CF	EF	GF	BF	AF	DF	0.6
FG	IG	HG	CG	EG	GG	BG	AG	DG	0.7
FH	IH	HH	CH	EH	GH	BH	AH	DH	0.8
FI	II	HI	CI	EI	GI	BI	AI	DI	0.9

*水澤腹堅。 Streams and marshes are [frozen] hard to their bottoms.

Comment: In this time of Achievement Good Luck is attained when one acts with propriety and kindness. Still, the world outside is frozen solid, and there is nothing to be gained by leaving one's retreat. This is not yet time to lead a new venture. If by doing so one spoils what one hoped to achieve, that is Bad Luck.

冲 二之二

冲。元亨。大君體仁。首出庶物；萬國以寧。無不利。

Peace[19] *Chong* 2.2

Peace. Prodigious Success. The great superior embodies benevolence.[20] The leader stands out above the many beings; the myriad countries because of it are at peace.[21] Nothing does not profit.

數曰:

The Numbers say:

9	8	7	6	5	4	3	2	1	
FD	ID	HD	CD	ED	GD	BD	AD	DD	0.1
FA	IA	HA	CA	EA	GA	BA	AA*	DA	0.2
FB	IB	HB	CB	EB	GB	BB	AB	DB	0.3
FG	IG	HG	CG	EG	GG	BG	AG	DG	0.4
FE	IE	HE	CE	EE	GE	BE	AE	DE	0.5
FC	IC	HC	CC	EC	GC	BC	AC	DC	0.6
FH	IH	HH	CH	EH	GH	BH	AH	DH	0.7
FI	II	HI	CI	EI	GI	BI	AI	DI	0.8
FF	IF	HF	CF	EF	GF	BF	AF	DF	0.9

*立春；東風解凍。 Advent of Spring;[22] east wind breaks the freeze.

Comment: The new year begins with the promise of spring; the deep freeze is broken. One feels Peace and knows unimpeded success. The noble-minded person acts in the spirit of benevolence to all and thereby emerges as leader of all, bringing peace far and wide. Every undertaking brings gains.

振 二至三

振。宣布文德。率作怠慢不恭，凶。

Arousal *Zhen* 2.3

Arousal. Pronounce and promulgate refined virtue. Carelessly display negligence and show no respect—Bad Luck.

數曰:

The Numbers say:

9	8	7	6	5	4	3	2	1	
FG	IG	HG	CG	EG	GG	BG	AG	DG	0.1
FD	ID	HD	CD	ED	GD	BD	AD	DD	0.2
FA	IA	HA	CA	EA	GA	BA*	AA	DA	0.3
FH	IH	HH	CH	EH	GH	BH	AH	DH	0.4
FE	IE	HE	CE	EE	GE	BE	AE	DE	0.5
FB	IB	HB	CB	EB	GB	BB	AB	DB	0.6
FI	II	HI	CI	EI	GI	BI	AI	DI	0.7
FF	IF	HF	CF	EF	GF	BF	AF	DF	0.8
FC	IC	HC	CC	EC	GC	BC	AC	DC	0.9

*蟄蟲始振。 Hibernating insects are first aroused.

Comment: The first stirrings of spring arouse the impulse to action. In response, one should urge others to practice the humanizing virtues taught by the sages and do so oneself. If, on the contrary, one is negligent and disrespectful, it will bring Bad Luck.

祈 二之四

祈。求而往。無不利。祭祀，吉

Prayer *Qi* 2.4

Prayer. Seek and go; nothing does not profit. Sacrifice, worship—Good Luck.

數曰:

The Numbers say:

9	8	7	6	5	4	3	2	1	
FB	IB	HB	CB	EB	GB	BB	AB	DB	0.1
FC	IC	HC	CC	EC	GC	BC	AC	DC	0.2
FF	IF	HF	CF	EF	GF	BF	AF	DF	0.3
FA	IA	HA	CA	EA	GA*	BA	AA	DA	0.4
FE	IE	HE	CE	EE	GE	BE	AE	DE	0.5
FI	II	HI	CI	EI	GI	BI	AI	DI	0.6
FD	ID	HD	CD	ED	GD	BD	AD	DD	0.7
FG	IG	HG	CG	EG	GG	BG	AG	DG	0.8
FH	IH	HH	CH	EH	GH	BH	AH	DH	0.9

*魚上冰。 Fish rise to the ice.

Comment: The time of Prayer is late winter when there are incipient signs of rebirth, but means of sustenance are difficult to find. If one seeks support from the spirits, one can go out looking and believe that one's chances will be favored. Thus, sacrifices and rituals bring Good Luck.

常 二之五

常。元亨。利不息之貞。[23]

Constancy *Chang* 2.5

Constancy. Prodigious Success. Profit in unceasing purity.

數曰:

The Numbers say:

9	8	7	6	5	4	3	2	1	
FI	IF	HC	CH	EE	GB	BG	AD	DA	0.1
FH	II	HF	CG	EE	GC	BD	AA	DB	0.2
FG	IH	HI	CD	EE	GF	AB	AB	DC	0.3
FF	IC	HB	CI	EE	GA	BH	AG	DD	0.4
FA	IA	HA	CA	EA*	GA	BA	AA	DA	0.5
FD	IG	HH	CA	EE	GI	BB	AC	DF	0.6
FC	IB	HA	CF	EE	GD	BI	AH	DG	0.7
FB	IA	HD	CC	EE	GG	BF	AI	DH	0.8
FA	ID	HG	CB	EE	GH	BC	AF	DI	0.9

*雨水；獺祭魚。 Rain Water;[24] otters make sacrifice of fish.

Comment: Constancy is a moment of prodigious success, of undertakings free of obstacle. One's actions will bring profit if one is guided by the innate sense of what is right undiluted by self-centered distractions. Then one exhibits the selfless regularity of the four seasons and the phases of the moon, which occur with unceasing purity in their proper times.

柔 二之六

柔。惠。利用正。婦人吉。夫子，凶。

Softness *Rou* 2.6

Softness. Kindness. Profit in enacting propriety. Wife—Good Luck. Husband—Bad Luck.[25]

數曰:

The Numbers say:

9	8	7	6	5	4	3	2	1	
FH	IH	HH	CH	EH	GH	BH	AH	DH	0.1
FG	IG	HG	CG	EG	GG	BG	AG	DG	0.2
FD	ID	HD	CD	ED	GD	BD	AD	DD	0.3
FI	II	HI	CI	EI	GI	BI	AI	DI	0.4
FE	IE	HE	CE	EE	GE	BE	AE	DE	0.5
FA	IA	HA	CA*	EA	GA	BA	AA	DA	0.6
FF	IF	HF	CF	EF	GF	BF	AF	DF	0.7
FC	IC	HC	CC	EC	GC	BC	AC	DC	0.8
FB	IB	HB	CB	EB	GB	BB	AB	DB	0.9

*鴻雁來。 Swans and geese come.[26]

Comment: The Softness of spring is the opposite of the hard icy winter past and to come. Softness is kind, yet remains within the bounds of propriety and thereby gains advantage. Softness is the way of a woman and wife, and in a relationship it now brings Good Luck. But the masculine rigidity that a husband might display will result in Bad Luck.

易 二之七

易。百物順生；庶事順；平易近民。[27] 艱險凶；不利涉大川。

Tranquility[28] *Yi* 2.7

Tranquility. The hundred things are produced in order; the many matters are achieved in order. Balance and tranquility bring one close to the people. Difficulty and adversity—Bad Luck. No profit in crossing the great river.

數曰:

The Numbers say:

9	8	7	6	5	4	3	2	1	
FC	IC	HC	CC	EC	GC	BC	AC	DC	0.1
FF	IF	HF	CF	EF	GF	BF	AF	DF	0.2
FI	II	HI	CI	EI	GI	BI	AI	DI	0.3
FB	IB	HB	CB	EB	GB	BB	AB	DB	0.4
FE	IE	HE	CE	EE	GE	BE	AE	DE	0.5
FH	IH	HH	CH	EH	GH	BH	AH	DH	0.6
FA	IA	HA*	CA	EA	GA	BA	AA	DA	0.7
FD	ID	HD	CD	ED	GD	BD	AD	DD	0.8
FG	IG	HG	CG	EG	GG	BG	AG	DG	0.9

*草木萌動。 Sprouts of grasses and trees move.

Comment: A time of Tranquility in the environment requires inner tranquility. It can be found through understanding the orderly progress that underlies the natural round. Living things come into being each in their season—at this moment it is the first shoots from the soil. And events likewise follow a general course, beginning now and proceeding to grow. The balanced perspective that is manifest in tranquility makes it possible to work with the people, to lead them if one is a superior. But it is a delicate balance, and the stress of difficult adversity will tip one into Bad Luck. No gain will be had if one undertakes the risk of crossing the great river.

親 二之八

親。內和順，而外文明。父父、子子、兄兄、弟弟、夫夫、婦婦。上下睦，而家道亨。

Kinship *Qin* 2.8

Kinship. Inside is harmonious order, and cultivation shines brightly on the outside.[29] The father is a father, the child a child, elder sibling is elder sibling, younger is younger, husband is husband, and wife is wife. Above and below there is accord and the Way of the family has Success.

數曰:

The Numbers say:

9	8	7	6	5	4	3	2	1	
FF	IF	HF	CF	EF	GF	BF	AF	DF	0.1
FI	II	HI	CI	EI	GI	BI	AI	DI	0.2
FH	IH	HH	CH	EH	GH	BH	AH	DH	0.3
FC	IC	HC	CC	EC	GC	BC	AC	DC	0.4
FE	IE	HE	CE	EE	GE	BE	AE	DE	0.5
FG	IG	HG	CG	EG	GG	BG	AG	DG	0.6
FB	IB	HB	CB	EB	GB	BB	AB	DB	0.7
FA	IA*	HA	CA	EA	GA	BA	AA	DA	0.8
FD	ID	HD	CD	ED	GD	BD	AD	DD	0.9

*驚蟄；桃始華。 Waking of Insects;[30] peach first blossoms.

Comment: Kinship is a time when the almanac says life becomes visible in the appearance of insects and peach blossoms. The family is harmonious and orderly among themselves, and to the outer world it radiates ideal behavior. The father is a true father in the best sense of the word, supporting, instructing, disciplining. The child is filial and obedient, as a child should be; elder siblings are generous to the younger; so also other family members fulfill their expectations. Within the home, all know and perform their interrelated duties, so that from its higher to its lower levels, everything is seen to be in accord. Thus, fulfilling the Way of the family results in success.

華 二之九

華。文明以正。利有攸行。勿用折獄。木道，[31] 亨。

Flowering *Hua* 2.9

Flowering. Through propriety one's cultivation radiates. Profit in taking action. Do not judge lawsuits. The Way of wood—Success.

數曰:

The Numbers say:

9	8	7	6	5	4	3	2	1	
FI	II	HI	CI	EI	GI	BI	AI	DI	0.1
FH	IH	HH	CH	EH	GH	BH	AH	DH	0.2
FG	IG	HG	CG	EG	GG	BG	AG	DG	0.3
FF	IF	HF	CF	EF	GF	BF	AF	DF	0.4
FE	IE	HE	CE	EE	GE	BE	AE	DE	0.5
FD	ID	HD	CD	ED	GD	BD	AD	DD	0.6
FC	IC	HC	CC	EC	GC	BC	AC	DC	0.7
FB	IB	HB	CB	EB	GB	BB	AB	DB	0.8
FA	IA	HA	CA	EA	GA	BA	AA	DA	0.9

Comment: The blossoms that first appeared in the previous period come to full bloom in the time of Flowering. Now righteousness—which means suiting one's actions to one's true nature—brings one's cultivation into brilliant display. Now there is profit in taking action, though legal action is unadvisable. Spring is the season of wood—of the growth of stems and leaves from trunks that for months seemed lacking in life, of seeds that have lain in wait. Renewal is the Way of wood, and it brings success.

見 三之一

見。一氣既信；百有著形。睟面，盎背；德潤厥身。隱匿，凶。

Seeing *Jian* 3.1

Visibility. Elemental unity has extended; it is manifest in hundreds of shapes. Clear-eyed the face, broad the back;[32] virtue glistens over your body. Reclusive concealment—Bad Luck.

數曰:

The Numbers say:

9	8	7	6	5	4	3	2	1	
CA	FA	IA	BA	EA	HA	AA	DA	GA*	0.1
CB	FB	IB	BB	EB	HB	AB	DB	GB	0.2
CC	FC	IC	BC	EC	HC	AC	DC	GC	0.3
CD	FD	ID	BD	ED	HD	AD	DD	GD	0.4
CE	FE	IE	BE	EE	HE	AE	DE	GE	0.5
CF	FF	IF	BF	EF	HF	AF	DF	GF	0.6
CG	FG	IG	BG	EG	HG	AG	DG	GG	0.7
CH	FH	IH	BH	EH	HH	AH	DH	GH	0.8
CI	FI	II	BI	EI	HI	AI	DI	GI	0.9

*倉庚鳴。 Oriole sings.

Comment: Visibility comes with the elaboration of elemental *Qi*—the single medium that cycles between free energy and physical phenomena—into all its hundreds of shapes and forms. An individual's *Qi* reflects One's inner virtues; they show through in clear eyes and a straight, broad back. It anoints the whole body with a rich glow. Now is the time to appear in the world; hiding in reclusion will bring Bad Luck.

獲 三之二

獲。氣質形色；自天有得。君子遷善；小人革面。縱逸，凶。

Taking *Huo* 3.2

Taking. Elemental substance, its shapes and appearances—these are gotten from Heaven. Noble-minded person evolves to the good;[33] small-minded person alters his countenance.[34] Indulge in idleness—Bad Luck.

數曰:

The Numbers say:

9	8	7	6	5	4	3	2	1	
CD	FD	ID	BD	ED	HD	AD	DD	GD	0.1
CA	FA	IA	BA	EA	HA	AA	DA*	GA	0.2
CB	FB	IB	BB	EB	HB	AB	DB	GB	0.3
CG	FG	IG	BG	EG	HG	AG	DG	GG	0.4
CE	FE	IE	BE	EE	HE	AE	DE	GE	0.5
CC	FC	IC	BC	EC	HC	AC	DC	GC	0.6
CH	FH	IH	BH	EH	HH	AH	DH	GH	0.7
CI	FI	II	BI	EI	HI	AI	DI	GI	0.8
CF	FF	IF	BF	EF	HF	AF	DF	GF	0.9

*鷹化為鳩。 Hawks are transformed into doves.

Comment: As the spring equinox approaches, the early output of nature can be taken in the hunt and gathered in field and forest. These are all the elemental *Qi* that has taken the various shapes and appearances, and their templates are gotten from the creative force of Heaven. The noble-minded person is in awe of these dynamics underlying material existence and responds by turning always toward improvement. The small-minded person merely grins as if he understands. If one chooses to remain idle during so productive a time, it brings Bad Luck.

從 三之三

從。惟從非同。不獲其身；不見其人。利有攸行。

Following *Cong* 3.3

Following. The act of following[35] does not mean agreement. Does not take one's body; does not see one's person.[36] Profit in taking action.

數曰:

The Numbers say:

9	8	7	6	5	4	3	2	1	
CG	FG	IG	BG	EG	HG	AG	DG	GG	0.1
CD	FD	ID	BD	ED	HD	AD	DD	GD	0.2
CA	FA	IA	BA	EA	HA	AA*	DA	GA	0.3
CH	FH	IH	BH	EH	HH	AH	DH	GH	0.4
CE	FE	IE	BE	EE	HE	AE	DE	GE	0.5
CB	FB	IB	BB	EB	HB	AB	DB	GB	0.6
CI	FI	II	BI	EI	HI	AI	DI	GI	0.7
CF	FF	IF	BF	EF	HF	AF	DF	GF	0.8
CC	FC	IC	BC	EC	HC	AC	DC	GC	0.9

*春分；乙鳥[37]至。 Spring Equinox;[38] swallow arrives.

Comment: Daytime and nighttime are balanced at the spring equinox; similarly, one must be unbiased in judgement. Following an argument in a case and judging it to be true doesn't mean one must agree with the motivations of the person making the argument. This can feel elusive—like not being able to take hold of someone because you can't see him. But there is profit in taking action consistent with balanced thinking.

交 三之四

交。唱而和；感而應。渙汗大號。東南得朋。征伐，小利。

Mating *Jiao* 3.4

Mating. Sing and harmonize;[39] stimulate and respond.[40] Streaming sweat, give a great shout.[41] In the southeast, get a friend.[42] Attack, strike—Small Profit.

數曰:

The Numbers say:

9	8	7	6	5	4	3	2	1	
CB	FB	IB	BB	EB	HB	AB	DB	GB	0.1
CC	FC	IC	BC	EC	HC	AC	DC	GC	0.2
CF	FF	IF	BF	EF	HF	AF	DF	GF	0.3
CA	FA	IA	BA	EA	HA*	AA	DA	GA	0.4
CE	FE	IE	BE	EE	HE	AE	DE	GE	0.5
CI	FI	II	BI	EI	HI	AI	DI	GI	0.6
CD	FD	ID	BD	ED	HD	AD	DD	GD	0.7
CG	FG	IG	BG	EG	HG	AG	DG	GG	0.8
CH	FH	IH	BH	EH	HH	AH	DH	GH	0.9

*雷乃發聲。 Thunder now makes its sound.

Comment: Mating is the propensity of springtime. One sings; the other harmonizes. One stimulates; another responds. One streams with sweat, then gives a great shout of consummation. East on the compass is spring on the yearly round, and now the time is just south of east when thunder first rouses the emotions. Here one finds the friend with whom one will interact. With this friend to assist, there is some small profit to be gained from launching an attack.

育 三之五

育。天地絪縕。萬物化醇。聖人順成。生產，吉。

Nurturing *Yu* 3.5

Nurturing. Heaven and earth intertwine threads; the myriad things transform, each to its essence.[43] The sage achieves in orderly fashion. Giving birth—Good Luck.

數曰:

The Numbers say:

9	8	7	6	5	4	3	2	1	
CI	FF	IC	BH	EE	HB	AG	DD	GA	0.1
CH	FI	IF	BG	EE	HC	AD	DA	GB	0.2
CG	FH	II	BD	EE	HF	AA	DB	GC	0.3
CF	FC	IB	BI	EE	HA	AH	DG	GD	0.4
CA	FA	IA	BA	EA*	HA	AA	DA	GA	0.5
CD	FG	IH	BA	EE	HI	AB	DC	GF	0.6
CC	FB	IA	BF	EE	HD	AI	DH	GG	0.7
CB	FA	ID	BC	EE	HG	AF	DI	GH	0.8
CA	FD	IG	BB	EE	HH	AC	DF	GI	0.9

*始電。 First lightning.

Comment: Nurturing follows Mating, as the feminine and masculine powers of nature intertwine to form new life, and the myriad things reproduce according to their kind. The sage follows a similar orderly progression and achieves the objective in season. Giving birth in this time is Good Luck.

壯 三之六

壯于正。有攸往，無不利。

Strength *Zhuang* 3.6

Strength in propriety. There is somewhere to go—nothing does not profit.

數曰:

The Numbers say:

9	8	7	6	5	4	3	2	1	
CH	FH	IH	BH	EH	HH	AH	DH	GH	0.1
CG	FG	IG	BG	EG	HG	AG	DG	GG	0.2
CD	FD	ID	BD	ED	HD	AD	DD	GD	0.3
CI	FI	II	BI	EI	HI	AI	DI	GI	0.4
CE	FE	IE	BE	EE	HE	AE	DE	GE	0.5
CA	FA	IA	BA*	EA	HA	AA	DA	GA	0.6
CF	FF	IF	BF	EF	HF	AF	DF	GF	0.7
CC	FC	IC	BC	EC	HC	AC	DC	GC	0.8
CB	FB	IB	BB	EB	HB	AB	DB	GB	0.9

*清明；桐始華。 Clear Bright:[44] tung tree first blossoms.

Comment: One is strong, and that Strength is nurtured by propriety. One can go out and do anything. There is nothing that will not bring profit.

興 三之七

興。吉。利見大人。天下文明。萬邦黎獻方來不寧。土役無度，凶。

Uplifting *Xing* 3.7

Uplifting. Good Luck. Profit in seeing the great person. The designs beneath heaven are brilliantly displayed.[45] From the myriad lands the worthies among the common people[46] come from all directions, not at a leisurely pace. If local service is not measured—Bad Luck.

數曰:

The Numbers say:

9	8	7	6	5	4	3	2	1	
CC	FC	IC	BC	EC	HC	AC	DC	GC	0.1
CF	FF	IF	BF	EF	HF	AF	DF	GF	0.2
CI	FI	II	BI	EI	HI	AI	DI	GI	0.3
CB	FB	IB	BB	EB	HB	AB	DB	GB	0.4
CE	FE	IE	BE	EE	HE	AE	DE	GE	0.5
CH	FH	IH	BH	EH	HH	AH	DH	GH	0.6
CA	FA	IA*	BA	EA	HA	AA	DA	GA	0.7
CD	FD	ID	BD	ED	HD	AD	DD	GD	0.8
CG	FG	IG	BG	EG	HG	AG	DG	GG	0.9

*田鼠化為鴽。 Moles are transformed into quail.

Comment: Uplifting is a time of Good Luck, a time when it is beneficial to learn from those of proven worth. The innate patterns of growth and interrelationship among all beings under heaven are clearly visible. The most knowledgeable of the common folk hurry to one's presence from all the surrounding countries to offer advice. If the demands for service to the state put on the people are not kept in check, those worthies will not come and their compatriots will become mutinous. That would be Bad Luck.

欣 三之八

欣。氣和時評。[47]萬物向榮。君子樂道；小人樂生。淫於酒，喪其明，凶。

Gladness *Xin* 3.8

Gladness. Elements are in harmony; times are peaceful. The myriad things tend to flourishing. Noble-minded person takes joy in the Way, small-minded person takes joy in being alive. Overindulging in wine, losing one's clarity—Bad Luck.

數曰:

The Numbers say:

9	8	7	6	5	4	3	2	1	
CF	FF	IF	BF	EF	HF	AF	DF	GF	0.1
CI	FI	II	BI	EI	HI	AI	DI	GI	0.2
CH	FH	IH	BH	EH	HH	AH	DH	GH	0.3
CC	FC	IC	BC	EC	HC	AC	DC	GC	0.4
CE	FE	IE	BE	EE	HE	AE	DE	GE	0.5
CG	FG	IG	BG	EG	HG	AG	DG	GG	0.6
CB	FB	IB	BB	EB	HB	AB	DB	GB	0.7
CA	FA*	IA	BA	EA	HA	AA	DA	GA	0.8
CD	FD	ID	BD	ED	HD	AD	DD	GD	0.9

*虹始見。 Rainbow first seen.

Comment: Be Glad: the elements are in harmony, it is a time of peace, and the first rainbow hovers in a warm spring sky. Everything flourishes. The noble-minded person finds joy in understanding the patterned Way of nature; the small-minded person is happy just to be alive. But one should be moderate in celebrating—overindulging in wine and losing one's clarity of mind brings Bad Luck.

舒 三之九

舒。雨露霑濡R；草木榮敷R。百體以舒R；惟仁之腴R。無不利。迫近，凶。

Soothing *Shu* 3.9

Soothing. Rain, dew moisten, soak; grass, wood are in flourishing array. The hundred forms avail of Soothing; acts of benevolence arrive at richness. There is nothing that does not profit. Forced nearness—Bad Luck.

數曰:

The Numbers say:

9	8	7	6	5	4	3	2	1	
CI	FI	II	BI	EI	HI	AI	DI	GI	0.1
CH	FH	IH	BH	EH	HH	AH	DH	GH	0.2
CG	FG	IG	BG	EG	HG	AG	DG	GG	0.3
CF	FF	IF	BF	EF	HF	AF	DF	GF	0.4
CE	FE	IE	BE	EE	HE	AE	DE	GE	0.5
CD	FD	ID	BD	ED	HD	AD	DD	GD	0.6
CC	FC	IC	BC	EC	HC	AC	DC	GC	0.7
CB	FB	IB	BB	EB	HB	AB	DB	GB	0.8
CA	FA	IA	BA	EA	HA	AA	DA	GA	0.9

Comment: In the time of Soothing, timely showers and morning dews moisten the flourishing young vegetation in fields and meadows. All types of beings are thereby soothed, and acts of benevolence atop what comes naturally make for richness. There is nothing that does not profit. But when all can enjoy what is freely available, forcing things together brings Bad Luck.

比 四之一

比。上下相親[R]；左右承鄰[R]。龍見雲升。君子以衆[R2]；小人勿用[R2]。

Nearness *Bi* 4.1

Nearness. Above and below are like kin; left and right are neighborly. Dragons see clouds and ascend. Noble-minded person utilizes the masses. Small-minded person must not be employed.

數曰:

The Numbers say:

9	8	7	6	5	4	3	2	1	
HA	GA	DA	IA	EA	AA	FA	CA	BA*	0.1
HB	GB	DB	IB	EB	AB	FB	CB	BB	0.2
HC	GC	DC	IC	EC	AC	FC	CC	BC	0.3
HD	GD	DD	ID	ED	AD	FD	CD	BD	0.4
HE	GE	DE	IE	EE	AE	FE	CE	BE	0.5
HF	GF	DF	IF	EF	AF	FF	CF	BF	0.6
HG	GG	DG	IG	EG	AG	FG	CG	BG	0.7
HH	GH	DH	IH	EH	AH	FH	CH	BH	0.8
HI	GI	DI	II	EI	AI	FI	CI	BI	0.9

*穀雨；萍始生。 Grain Rain;[48] duckweed first grows.

Comment: The spring rains fall; people join together in Nearness. Regardless of station, high or low, they are like relations and neighborly to each other. The dragons, which have emerged from their winter hibernation, see the clouds and ascend to them. The noble-minded person guides masses of people in their work, but it is not a time to employ small-minded people.

開 四之二

開。析民墾田。闢塞通障。利有攸往。閉糶藏埜，凶。

Opening *Kai* 4.2

Opening. Organize the people to cultivate the fields. Clear blockages and penetrate hindrances. Profit in going somewhere. Curtailing sales, storing, and burying—Bad Luck.

數曰:

The Numbers say:

9	8	7	6	5	4	3	2	1	
HD	GD	DD	ID	ED	AD	FD	CD	BD	0.1
HA	GA	DA	IA	EA	AA	FA	CA*	BA	0.2
HB	GB	DB	IB	EB	AB	FB	CB	BB	0.3
HG	GG	DG	IG	EG	AG	FG	CG	BG	0.4
HE	GE	DE	IE	EE	AE	FE	CE	BE	0.5
HC	GC	DC	IC	EC	AC	FC	CC	BC	0.6
HH	GH	DH	IH	EH	AH	FH	CH	BH	0.7
HI	GI	DI	II	EI	AI	FI	CI	BI	0.8
HF	GF	DF	IF	EF	AF	FF	CF	BF	

*鳴鳩拂其羽。 Cooing doves flutter their wings.

Comment: In the time of Opening, the danger of killing frost is passing, and one can direct the people to begin farming. They can clear blockages from irrigation channels, dig up stumps and stones, and break through any obstacle left over from winter storms. There is profit to be gained from going out and seeking it. If instead one were to keep markets closed or store things away in the ground, it would bring Bad Luck.

晉 四之三

晉。進賢去邪；百工咸理。監工日號。悖于時，凶。蠶桑，吉。

Progress *Jin* 4.3

Progress. Draw near to worthies; cast out the perverse. The hundred kinds of work are all managed. In supervising work, daily shout encouragement. Contrary to the times—Bad Luck. Nurturing silkworms and mulberry—Good Luck.

數曰:

The Numbers say:

9	8	7	6	5	4	3	2	1	
HG	GG	DG	IG	EG	AG	FG	CG	BG	0.1
HD	GD	DD	ID	ED	AD	FD	CD	BD	0.2
HA	GA	DA	IA	EA	AA	FA*	CA	BA	0.3
HH	GH	DH	IH	EH	AH	FH	CH	BH	0.4
HE	GE	DE	IE	EE	AE	FE	CE	BE	0.5
HB	GB	DB	IB	EB	AB	FB	CB	BB	0.6
HI	GI	DI	II	EI	AI	FI	CI	BI	0.7
HF	GF	DF	IF	EF	AF	FF	CF	BF	0.8
HC	GC	DC	IC	EC	AC	FC	CC	BC	0.9

*戴勝降於桑。 Hoopoes alight in the mulberry trees.

Comment: Progress is made. It is a time to depend on the worthiest people and shunt aside the troublesome ones. Manage all the various tasks that need to be done, and as you do, shout out encouragement to your workers every day. If you act counter to the positive energies of the moment, it will bring Bad Luck. There is Good Luck in the regular tasks of husbandry, like tending to the silkworm cocoons and the mulberry trees whose leaves will feed them.

公 四之四

公。亨。天高地下。萬物散殊[R]。君子克己；禮復其初[R]。利折獄。

Altruism *Gong* 4.4

Altruism. Success. Heaven is high, earth below. Myriad things are distributed in all their diversity.[49] Noble-minded person constrains ego; in rites returns[50] to one's beginning. Profit in judging lawsuits.

數曰:

The Numbers say:

9	8	7	6	5	4	3	2	1	
HB	GB	DB	IB	EB	AB	FB	CB	BB	0.1
HC	GC	DC	IC	EC	AC	FC	CC	BC	0.2
HF	GF	DF	IF	EF	AF	FF	CF	BF	0.3
HA	GA	DA	IA	EA	AA*	FA	CA	BA	0.4
HE	GE	DE	IE	EE	AE	FE	CE	BE	0.5
HI	GI	DI	II	EI	AI	FI	CI	BI	0.6
HD	GD	DD	ID	ED	AD	FD	CD	BD	0.7
HG	GG	DG	IG	EG	AG	FG	CG	BG	0.8
HH	GH	DH	IH	EH	AH	FH	CH	BH	0.9

*立夏；螻蟈鳴。 Advent of Summer;[51] mole crickets chirrup.

Comment: Public spirited Altruism reigns in anticipation of summer's energies. Activity brings Success. Heaven is high, earth is below; the myriad things are distributed between in all their diversity. Feeling one's powers surging, the noble-minded person curbs the ego by submitting to the discipline of time-honored ritual. Rituals at first are learned by rote, and though one comes to appreciate their deeper significance as time goes by, one is transported back to the ingenuous obedience of the first-time learner when performing them. One gains from judging lawsuits.

益 四之五

益。友朋方來敬之。終，吉。繼長增高；與時偕極。廢惰，凶。

Increase *Yi* 4.5

Increase. Companions come from the [four] quarters and honor one. Ending—Good Luck. Continuing growth, adding to elevation; things with time together reach their peak. Rejected, indolent—Bad Luck.

數曰:

The Numbers say:

9	8	7	6	5	4	3	2	1	
HI	GF	DC	IH	EE	AB	FG	CD	BA	0.1
HH	GI	DF	IG	EE	AC	FD	CA	BB	0.2
HG	GH	DI	ID	EE	AF	FA	CB	BC	0.3
HF	GC	DB	II	EE	AA	FH	CG	BD	0.4
HA	GA	DA	IA	EA*	AA	FA	CA	BA	0.5
HD	GG	DH	IA	EE	AI	FB	CC	BF	0.6
HC	GB	DA	IF	EE	AD	FI	CH	BG	0.7
HB	GA	DD	IC	EE	AG	FF	CI	BH	0.8
HA	GD	DG	IB	EE	AH	FC	CF	BI	0.9

*蚯蚓出。 Earthworms emerge.

Comment: The time of Increase sees friends come from all around and pay homage. This culminates in Good Luck. The summer's growth goes on, taking the crops and one's own fortunes higher and higher until all reaches its peak. If one is not employed or lazy, it will bring Bad Luck.

章 四之六

章。天下文明；赫赫彬彬。大震厥聲。匪正，有悔。

Distinctness[52] *Zhang* 4.6

Distinctness. The designs beneath heaven are brilliantly displayed. Illustrious, illustrious; elegant, elegant. Great thunder in your voice. Lacking propriety—have regret.

數曰:

The Numbers say:

9	8	7	6	5	4	3	2	1	
HH	GH	DH	IH	EH	AH	FH	CH	BH	0.1
HG	GG	DG	IG	EG	AG	FG	CG	BG	0.2
HD	GD	DD	ID	ED	AD	FD	CD	BD	0.3
HI	GI	DI	II	EI	AI	FI	CI	BI	0.4
HE	GE	DE	IE	EE	AE	FE	CE	BE	0.5
HA	GA	DA	IA	EA*	AA	FA	CA	BA	0.6
HF	GF	DF	IF	EF	AF	FF	CF	BF	0.7
HC	GC	DC	IC	EC	AC	FC	CC	BC	0.8
HB	GB	DB	IB	EB	AB	FB	CB	BB	0.9

*王瓜生。 Royal melon sprouts.

Comment: In Distinctness, the natural designs of all beings and processes are brilliantly displayed. They are illustrious and elegant. Your voice rings out with power and authority, like the sound of thunder. But people will remember if your tone or words lack propriety, and you will regret it.

盈 四之七

盈。生氣流形R；品物咸亨R。雷雨滿盈R；不疑其行R。

Fullness *Ying* 4.7

Fullness. Elements of life flow through shapes;[53] the many types of things are all successful.[54] Thunder, rain make fullness replete.[55] No doubt in one's action.

數曰:

The Numbers say:

9	8	7	6	5	4	3	2	1	
HC	GC	DC	IC	EC	AC	FC	CC	BC	0.1
HF	GF	DF	IF	EF	AF	FF	CF	BF	0.2
HI	GI	DI	II	EI	AI	FI	CI	BI	0.3
HB	GB	DB	IB	EB	AB	FB	CB	BB	0.4
HE	GE	DE	IE	EE	AE	FE	CE	BE	0.5
HH	GH	DH	IH	EH	AH	FH	CH	BH	0.6
HA	GA	DA*	IA	EA	AA	FA	CA	BA	0.7
HD	GD	DD	ID	ED	AD	FD	CD	BD	0.8
HG	GG	DG	IG	EG	AG	FG	CG	BG	0.9

*小滿；苦菜秀。 Small Abundance;[56] sow thistle blooms.

Comment: Fullness means that as the living *Qi* energy flows through all the various forms it takes, each of the things it becomes now attains success. The thunder and rain of this time arouse and nourish them to the fullest. One has no doubts about the action one takes.

錫 四之八

錫。亨。厲。發爵賜服；慶賞以行。小人勿承以殃厥身。

Bestowing *Xi* 4.8

Bestowing. Success. Threat. Pass the cup [to install in a rank], grant a robe [as the insignia of office];[57] congratulate and award according to actions. Do not depend on a small-minded person without bringing calamity upon yourself.

數曰:

The Numbers say:

9	8	7	6	5	4	3	2	1	
HF	GF	DF	IF	EF	AF	FF	CF	BF	0.1
HI	GI	DI	II	EI	AI	FI	CI	BI	0.2
HH	GH	DH	IH	EH	AH	FH	CH	BH	0.3
HC	GC	DC	IC	EC	AC	FC	CC	BC	0.4
HE	GE	DE	IE	EE	AE	FE	CE	BE	0.5
HG	GG	DG	IG	EG	AG	FG	CG	BG	0.6
HB	GB	DB	IB	EB	AB	FB	CB	BB	0.7
HA	GA*	DA	IA	EA	AA	FA	CA	BA	0.8
HD	GD	DD	ID	ED	AD	FD	CD	BD	0.9

*靡草死。 Supple grasses[58] die.

Comment: Now that the cycle of nature is in its fullness, bestowing rewards on one's supporters brings Success. But there is also a threatening aspect to the moment. Following the old customs, pass the cup as a sign of installing people in positions of authority and grant ceremonial robes to identify their official roles. Make these gestures to recognize good works and congratulate those who do them. However, if you put small-minded people in positions of responsibility, you will bring disaster upon yourself.

靡 四之九

靡。亨。上下謐寧。來庭來賓，勿狥其名。大人，吉；小人，吝。疾病，凶。

Revelry[59] *Mi* 4.9

Revelry. Success. Those higher and lower are idle and at ease. They come to one's halls—come to sojourn, not seeking to enhance their reputations.[60] Great person—Good Luck. Small-minded person—remorse. Sickness, disease—Bad Luck.

數曰:

The Numbers say:

9	8	7	6	5	4	3	2	1	
HI	GI	DI	II	EI	AI	FI	CI	BI	0.1
HH	GH	DH	IH	EH	AH	FH	CH	BH	0.2
HG	GG	DG	IG	EG	AG	FG	CG	BG	0.3
HF	GF	DF	IF	EF	AF	FF	CF	BF	0.4
HE	GE	DE	IE	EE	AE	FE	CE	BE	0.5
HD	GD	DD	ID	ED	AD	FD	CD	BD	0.6
HC	GC	DC	IC	EC	AC	FC	CC	BC	0.7
HB	GB	DB	IB	EB	AB	FB	CB	BB	0.8
HA	GA	DA	IA	EA	AA	FA	CA	BA	0.9

Comment: The bounty of the season permits Revelry. All classes of people are idle and relaxed. They come to enjoy themselves in one's halls, putting aside their ambition to make names for themselves through their works to enjoy your company. For the mature, broad-minded person, this is a time of Good Luck; for the small-minded, it will lead to remorse. Sickness from overindulgence spells Bad Luck.

庶 五之一

庶。天開地闢；萬物蕃殖。君子所體：利衆，不利寡；利公，不利私。

Abundance *Shu* 5.1

Abundance. Heaven opens, earth bursts out;[61] myriad things breed and multiply. What the noble-minded person embodies is: profit the masses, do not profit the few; profit the public, do not profit private interests.

數曰:

The Numbers say:

9	8	7	6	5	4	3	2	1	
IA	HA	GA	FA	AA	DA	CA	BA	AA*	0.1
IB	HB	GB	FB	AB	DB	CB	BB	AB	0.2
IC	HC	GC	FC	AC	DC	CC	BC	AC	0.3
ID	HD	GD	FD	AD	DD	CD	BD	AD	0.4
IE	HE	GE	FE	AE	DE	CE	BE	AE	0.5
IF	HF	GF	FF	AF	DF	CF	BF	AF	0.6
IG	HG	GG	FG	AG	DG	CG	BG	AG	0.7
IH	HH	GH	FH	AH	DH	CH	BH	AH	0.8
II	HI	GI	FI	AI	DI	CI	BI	AI	0.9

*麥秋至。 Wheat attains ripeness.[62]

Comment: The numbers that mark the yearly round arrive at their central range; it is the time of Abundance beyond what has come before. Heaven and earth burst forth in bounty; the myriad plants and animals breed, pollinate, multiply. The stance of the noble-minded person is to bring this surfeit to the many, not horde it for the few; to profit the public, not selfish private interests.

決 五之二

決。八元舉用；四凶竄殛。羣疑盡釋；無枉不直。利艱正。

Decision *Jue* 5.2

Decision. Elevate for use the Eight Adepts;[63] expel and execute the Four Evil Ones.[64] The many doubts are entirely resolved. There is nothing crooked that will not be made straight. Profit in difficult propriety.

數曰:

The Numbers say:

9	8	7	6	5	4	3	2	1	
FD	ID	HD	CD	AD	GD	BD	AD	DD	0.1
FA	IA	HA	CA	AA	GA	BA	AA*	DA	0.2
FB	IB	HB	CB	AB	GB	BB	AB	DB	0.3
FG	IG	HG	CG	AG	GG	BG	AG	DG	0.4
FE	IE	HE	CE	AE	GE	BE	AE	DE	0.5
FC	IC	HC	CC	AC	GC	BC	AC	DC	0.6
FH	IH	HH	CH	AH	GH	BH	AH	DH	0.7
FI	II	HI	CI	AI	GI	BI	AI	DI	0.8
FF	IF	HF	CF	AF	GF	BF	AF	DF	0.9

*芒种；螳螂生。 Grain in Ear;[65] praying mantis born.

Comment: In the season of Abundance, when self-centered people seek to horde more than their share, one must show Decision. Like the legendary sage rulers, one employs the worthiest among one's followers to root out and punish the rapacious enemies of the common good. In this way, the many doubts about one's integrity are entirely resolved; everything that appeared crooked will be set straight. In this, propriety is difficult to put into action because it will stir enmity and even temporary strife; but it profits in the end.

豫 五之三

豫。飲食和樂。君子豫，吉。小人豫，凶。

Contentment *Yu* 5.3

Contentment. Drink, eat; harmony, joy. From the contentment of the noble-minded person comes Good Luck. From the contentment of the small-minded person comes Bad Luck.

數曰:

The Numbers say:

9	8	7	6	5	4	3	2	1	
CG	FG	IG	BG	AG	HG	AG	DG	GG	0.1
CD	FD	ID	BD	AD	HD	AD	DD	GD	0.2
CA	FA	IA	BA	AA	HA	AA*	DA	GA	0.3
CH	FH	IH	BH	AH	HH	AH	DH	GH	0.4
CE	FE	IE	BE	AE	HE	AE	DE	GE	0.5
CB	FB	IB	BB	AB	HB	AB	DB	GB	0.6
CI	FI	II	BI	AI	HI	AI	DI	GI	0.7
CF	FF	IF	BF	AF	HF	AF	DF	GF	0.8
CC	FC	IC	BC	AC	HC	AC	DC	GC	0.9

*鶌始鳴。 Cuckoo first calls.

Comment: Contentment comes when doubts are resolved and the prosperity of the season continues. Eat and drink in harmony and joy. In this, the contentment of the noble-minded person occasions Good Luck. But the contentment of the small-minded person, who tends to overindulgence, brings on Bad Luck.

升 五之四

升。禮明樂行R；萬化以成R。利見大人。不言有喻。允升，大吉。

Ascension *Sheng* 5.4

Ascension. Rites shed their brilliance; music is performed. The myriad transformations by these are completed. Profit in seeing the great person. Without words, have understanding. Confidently ascend—great Good Luck.

數曰:

The Numbers say:

9	8	7	6	5	4	3	2	1	
HB	GB	DB	IB	AB	AB	FB	CB	BB	0.1
HC	GC	DC	IC	AC	AC	FC	CC	BC	0.2
HF	GF	DF	IF	AF	AF	FF	CF	BF	0.3
HA	GA	DA	IA	AA	AA*	FA	CA	BA	0.4
HE	GE	DE	IE	AE	AE	FE	CE	BE	0.5
HI	GI	DI	II	AI	AI	FI	CI	BI	0.6
HD	GD	DD	ID	AD	AD	FD	CD	BD	0.7
HG	GG	DG	IG	AG	AG	FG	CG	BG	0.8
HH	GH	DH	IH	AH	AH	FH	CH	BH	0.9

*反舌無聲。 Blackbird makes no sound.

Comment: The time of Ascension comes just before the apex of solar energy at the summer solstice, the turning point that completes the first half of the yearly round. This fulfillment is captured in human terms by corresponding rituals and music. It is a time when, observing how the great man behaves in these ceremonies, one understands their meaning without need for verbal explanations. One may be confident in rising to the occasion. It is a time of great Good Luck.

中 五之五

中。赫赫大明[R]；耀彼四隣[R]。君子持盈[R]；小人毁成[R]。

Center *Zhong* 5.5

Center. Illustrious, illustrious; great light. Illuminate the four neighbors. Noble-minded person holds to fullness. Small-minded person destroys achievements.

數曰:

The Numbers say:

9	8	7	6	5	4	3	2	1	
EI	EF	EC	EH	AE	EB	EG	ED	EA	0.1
EH	EI	EF	EG	AE	EC	ED	EA	EB	0.2
EG	EH	EI	ED	AE	EF	EA	EB	EC	0.3
EF	EC	EB	EI	AE	EA	EH	EG	ED	0.4
EA	EA	EA	EA	AA*	EA	EA	EA	EA	0.5
ED	EG	EH	EA	AE	EI	EB	EC	EF	0.6
EC	EB	EA	EF	AE	ED	EI	EH	EG	0.7
EB	EA	ED	EC	AE	EG	EF	EI	EH	0.8
EA	ED	EG	EB	AE	EH	EC	EF	EI	0.9

*夏至；鹿角解。 Summer Solstice;[66] deer shed antlers.

Comment: Center marks the summer solstice, the day of maximum sunlight. Like the brilliance of the solstice sun at noon, its apex—and turning point toward darkness, one strives to enlighten those around in all directions. Thus centered, the noble-minded person, awed and humbled by the power of nature, is able to enjoy the bounty of fullness. The small-minded person, by contrast, will abuse the gifts of nature and undermine all that has been achieved.

伏 五之六

伏。不聞不覩；君子戒懼。勿用娶女。利潛師；不利有攸往。

Subservience *Fu* 5.6

Subservience. Not heard, not seen;[67] the noble-minded person is guarded, anxious. Do not take a mate. Profit in hiding the army. No profit in going anywhere.

數曰:

The Numbers say:

9	8	7	6	5	4	3	2	1	
BH	CH	FH	AH	AH	IH	DH	GH	HH	0.1
BG	CG	FG	AG	AG	IG	DG	GG	HG	0.2
BD	CD	FD	AD	AD	ID	DD	GD	HD	0.3
BI	CI	FI	AI	AI	II	DI	GI	HI	0.4
BE	CE	FE	AE	AE	IE	DE	GE	HE	0.5
BA	CA	FA	AA*	AA	IA	DA	GA	HA	0.6
BF	CF	FF	AF	AF	IF	DF	GF	HF	0.7
BC	CC	FC	AC	AC	IC	DC	GC	HC	0.8
BB	CB	FB	AB	AB	IB	DB	GB	HB	0.9

*蜩始鳴。 Cicada first chirrups.

Comment: The apex is just past, but one realizes the implication of the incipient trend toward darkness and decay. One feels one's Subservience to the round of nature and goes into retreat, being neither seen nor heard. In this and the next period, the noble-minded person feels anxious and defensive. This is no time to marry, and one will profit by secreting the army in a hidden encampment. There is no profit in going anywhere.

過 五之七

過。罔淫于樂。君子戒懼。君子過厚；小人過薄。利涉大川。

Excess *Guo* 5.7

Excess. Have no orgy of joy. Noble-minded person guarded, anxious.[68] The noble-minded person is excessively generous. The small-minded person is excessively miserly. Profit in fording the great river.

數曰:

The Numbers say:

9	8	7	6	5	4	3	2	1	
GC	DC	AC	HC	AC	BC	IC	FC	CC	0.1
GF	DF	AF	HF	AF	BF	IF	FF	CF	0.2
GI	DI	AI	HI	AI	BI	II	FI	CI	0.3
GB	DB	AB	HB	AB	BB	IB	FB	CB	0.4
GE	DE	AE	HE	AE	BE	IE	FE	CE	0.5
GH	DH	AH	HH	AH	BH	IH	FH	CH	0.6
GA	DA	AA*	HA	AA	BA	IA	FA	CA	0.7
GD	DD	AD	HD	AD	BD	ID	FD	CD	0.8
GG	DG	AG	HG	AG	BG	IG	FG	CG	0.9

*半夏生。 Crow-dipper sprouts.

Comment: What has been a season of plenty now feels like a time of Excess. This is no time for an orgy of joy—instead, the noble-minded person feels anxious and defensive, as in the preceding period. The noble-minded person is characterized by excessive generosity; the small-minded person is excessively miserly. Nonetheless, there is profit to be gained at this moment in undertaking a risky venture like crossing the great river.

疑 五之八

疑。有間有貳。君子用明；小人用罔。勿用決獄。凶。

Doubt *Yi* 5.8

Doubt. Having space between, there are two. Noble-minded person enlightens; small-minded person uses negativity. Do not decide lawsuits. Bad Luck.

數曰:

The Numbers say:

9	8	7	6	5	4	3	2	1	
DF	AF	BF	GF	AF	CF	HF	IF	FF	0.1
DI	AI	BI	GI	AI	CI	HI	II	FI	0.2
DH	AH	BH	GH	AH	CH	HH	IH	FH	0.3
DC	AC	BC	GC	AC	CC	HC	IC	FC	0.4
DE	AE	BE	GE	AE	CE	HE	IE	FE	0.5
DG	AG	BG	GG	AG	CG	HG	IG	FG	0.6
DB	AB	BB	GB	AB	CB	HB	IB	FB	0.7
DA	AA*	BA	GA	AA	CA	HA	IA	FA	0.8
DD	AD	BD	GD	AD	CD	HD	ID	FD	0.9

*小暑；溫風至。 Small Heat;[69] warm winds blow.

Comment: Doubts arise as the summer heat intensifies. Where there is any distance between them, people are of two minds rather than of one. The noble-minded person deals with doubts by enlightening people as to the true facts; the small-minded person amplifies doubts by being negative and querulous. Do not attempt to decide lawsuits. It is a time of Bad Luck.

寡 五之九

寡。宜上不宜下；宜少不宜衆。君子寡過。不利婚媾。

Minimizing *Gua* 5.9

Minimizing. Benefits that above, does not benefit that below.[70] Benefits the few, does not benefit the masses. The noble-minded person minimizes his errors.[71] No profit in marrying.

數曰:

The Numbers say:

9	8	7	6	5	4	3	2	1	
AI	BI	CI	DI	AI	FI	GI	HI	II	0.1
AH	BH	CH	DH	AH	FH	GH	HH	IH	0.2
AG	BG	CG	DG	AG	FG	GG	HG	IG	0.3
AF	BF	CF	DF	AF	FF	GF	HF	IF	0.4
AE	BE	CE	DE	AE	FE	GE	HE	IE	0.5
AD	BD	CD	DD	AD	FD	GD	HD	ID	0.6
AC	BC	CC	DC	AC	FC	GC	HC	IC	0.7
AB	BB	CB	DB	AB	FB	GB	HB	IB	0.8
AA	BA	CA	DA	AA	FA	GA	HA	IA	0.9

Comment: In Minimizing, the noble-minded person reflects the coming decline of nature's proliferation by economizing. This is beneficial to those in higher places rather than those in lower ones; thus, it necessarily benefits the few rather than the masses in the short run. In the long run, though, it is the right course of action for all concerned. The noble-minded person takes this general course while seeking to minimize personal errors. There is no profit in marrying.

飾 六之一

飾。華文郁郁。貌恭作肅。君子謹獨

Decoration *Shi* 6.1

Decoration. Florid designs, lush, lush. One's respectful demeanor is shown in gravity.[72] Noble-minded person is cautious when alone.[73]

數曰:

The Numbers say:

9	8	7	6	5	4	3	2	1	
BA	CA	FA	AA	EA	IA	DA	GA	HA*	0.1
BB	CB	FB	AB	EB	IB	DB	GB	HB	0.2
BC	CC	FC	AC	EC	IC	DC	GC	HC	0.3
BD	CD	FD	AD	ED	ID	DD	GD	HD	0.4
BE	CE	FE	AE	EE	IE	DE	GE	HE	0.5
BF	CF	FF	AF	EF	IF	DF	GF	HF	0.6
BG	CG	FG	AG	EG	IG	DG	GG	HG	0.7
BH	CH	FH	AH	EH	IH	DH	GH	HH	0.8
BI	CI	FI	AI	EI	II	DI	GI	HI	0.9

*蟋蜂居壁。Crickets live in the walls.

Comment: Decoration sees rich ornamentation in one's chambers and furnishings. But one's demeanor exhibits respect displayed in gravity. The noble-minded person is cautious in behavior even when unobserved by others.

戾 六之二

戾。厲。吉。曲能有誠[R]；君子克明[R]。

Perversity *Li* 6.2

Perversity. Threat. Good Luck. Where there is an inclination, it can be turned to sincerity.[74] Noble-minded person is able to be brilliant.[75]

數曰:

The Numbers say:

9	8	7	6	5	4	3	2	1	
BD	CD	FD	AD	ED	ID	DD	GD	HD	0.1
BA	CA	FA	AA	EA	IA	DA	GA*	HA	0.2
BB	CB	FB	AB	EB	IB	DB	GB	HB	0.3
BG	CG	FG	AG	EG	IG	DG	GG	HG	0.4
BE	CE	FE	AE	EE	IE	DE	GE	HE	0.5
BC	CC	FC	AC	EC	IC	DC	GC	HC	0.6
BH	CH	FH	AH	EH	IH	DH	GH	HH	0.7
BI	CI	FI	AI	EI	II	DI	GI	HI	0.8
BF	CF	FF	AF	EF	IF	DF	GF	HF	0.9

*鷹乃學習 Hawks now learn to flap their wings.

Comment: In Perversity is threat but also Good Luck. Where the inclination toward the good exists—even in one who lacks refinement and might ignore it—the budding sensitivity can grow into sincerity. The noble-minded person is always able to let positive inclinations shine through.

虛 六之三

虛。理明而通[R]。應物不窮[R]。徇慾惟凶[R]。不利争訟[R]。

Emptiness *Xu* 6.3

Emptiness. Pattern is brilliant and open. Respond to things without ceasing.[76] Following desires leads to Bad Luck. No profit in struggle, dispute.

數曰:

The Numbers say:

9	8	7	6	5	4	3	2	1	
BG	CG	FG	AG	EG	IG	DG	GG	HG	0.1
BD	CD	FD	AD	ED	ID	DD	GD	HD	0.2
BA	CA	FA	AA	EA	IA	DA	GA*	HA	0.3
BH	CH	FH	AH	EH	IH	DH	GH	HH	0.4
BE	CE	FE	AE	EE	IE	DE	GE	HE	0.5
BB	CB	FB	AB	EB	IB	DB	GB	HB	0.6
BI	CI	FI	AI	EI	II	DI	GI	HI	0.7
BF	CF	FF	AF	EF	IF	DF	GF	HF	0.8
BC	CC	FC	AC	EC	IC	DC	GC	HC	0.9

*大暑；腐草為螢。 Great Heat;[77] rotting grass makes fireflies.

Comment: Emptiness is selfless humility that accepts without resistance. The grand pattern of nature shines through freely and openly. Though it is the hottest time of the year, one rouses oneself to respond relentlessly to whatever occurs. Blindly following desires leads to Bad Luck; there is nothing to be gained from struggling with others and engaging in disputes.

昧 六之四

昧。幽人正，吉。闇而章，晦而明。不利折獄。

Dimness *Mei* 6.4

Dimness. Propriety of secluded person—Good Luck. What is murky becomes distinct;[78] what is obscure comes to light.[79] No profit judging lawsuits.

數曰:

The Numbers say:

9	8	7	6	5	4	3	2	1	
BB	CB	FB	AB	EB	IB	DB	GB	HB	0.1
BC	CC	FC	AC	EC	IC	DC	GC	HC	0.2
BF	CF	FF	AF	EF	IF	DF	GF	HF	0.3
BA	CA	FA	AA	EA	IA*	DA	GA	HA	0.4
BE	CE	FE	AE	EE	IE	DE	GE	HE	0.5
BI	CI	FI	AI	EI	II	DI	GI	HI	0.6
BD	CD	FD	AD	ED	ID	DD	GD	HD	0.7
BG	CG	FG	AG	EG	IG	DG	GG	HG	0.8
BH	CH	FH	AH	EH	IH	DH	GH	HH	0.9

*土潤溽暑。 Soil is moist in the humid heat.

Comment: Though in high summer, there is a time of Dimness—the shadow of a dark cloud, the low light of a cool room. It is a time when the propriety of one living in seclusion brings Good Luck. Know that what is now murky will later become clear and distinct, that what is obscure will at some point come to light. This is not a time when advantage is gained from judging lawsuits.

損 六之五

損。君子之過日以削；小人之性日以斲。遇雨，吉。藥餌有喜。

Decrease *Sun* 6.5

Decrease. The noble-minded person daily pares away his errors; the small-minded person daily hacks off what arises from his innate nature. Meet with rain—Good Luck.[80] Medicine sweetener brings happiness.

數曰:

The Numbers say:

9	8	7	6	5	4	3	2	1	
BI	CF	FC	AH	EE	IB	DG	GD	HA	0.1
BH	CI	FF	AG	EE	IC	DD	GA	HB	0.2
BG	CH	FI	AD	EE	IH	DA	GB	HC	0.3
BF	CC	FB	AI	EE	IA	DH	GG	HD	0.4
BA	CA	FA	AA	EA*	IA	DA	GA	HA	0.5
BD	CG	FH	AA	EE	II	DB	GC	HF	0.6
BC	CB	FA	AF	EE	ID	DI	GH	HG	0.7
BB	CA	FD	AC	EE	IG	DF	GI	HH	0.8
BA	CD	FG	AB	EE	IH	DC	GF	HI	0.9

*大雨時行。 Great rains chance to come.

Comment: There is Decrease—new growth slackens as the crops ripen. The noble-minded person constantly whittles away the tendency to err; the small-minded person daily cuts off the tendency to good that arises from everyone's true, but often obscured nature. The almanac says it is a time when great rains chance to come; when one meets with rain, rather than dry spells that can as often occur, it is Good Luck.

用 六之六

用。利正。有攸往，吉。君子喻義；小人喻利。征伐有功。利决獄。

Utility *Yong* 6.6

Utility. Profit in propriety. Go somewhere—Good Luck. Noble-minded person thinks in terms of morality. Small-minded person thinks in terms of profit.[81] Attacking, striking bring victory. Profit in deciding lawsuits.

數曰:

The Numbers say:

9	8	7	6	5	4	3	2	1	
BH	CH	FH	AH	EH	IH	DH	GH	HH	0.1
BG	CG	FG	AG	EG	IG	DG	GG	HG	0.2
BD	CD	FD	AD	ED	ID	DD	GD	HD	0.3
BI	CI	FI	AI	EI	II	DI	GI	HI	0.4
BE	CE	FE	AE	EE	IE	DE	GE	HE	0.5
BA	CA	FA	AA*	EA	IA	DA	GA	HA	0.6
BF	CF	FF	AF	EF	IF	DF	GF	HF	0.7
BC	CC	FC	AC	EC	IC	DC	GC	HC	0.8
BB	CB	FB	AB	EB	IB	DB	GB	HB	0.9

*立秋；涼風至。 Advent of Autumn;[82] cool breezes arrive.

Comment: The advent of autumn turns the thoughts to Utility. There is profit from actions taken with propriety, and it is Good Luck to venture out. The noble-minded person evaluates utility in terms of morality; the small-minded person cares only for aggrandizement. Aggressive action—striking first—wins victory. It profits to decide lawsuits.

郤 六之七

郤。利行遯。反身以誠。不利有攸往。降責勿恤。

Withdrawal[83] *Shi* 6.7

Withdrawal. Profit enacting retreat. Examine oneself with sincerity. No profit in going anywhere. If demoted, do not worry.

數曰:

The Numbers say:

9	8	7	6	5	4	3	2	1	
BC	CC	FC	AC	EC	IC	DC	GC	HC	0.1
BF	CF	FF	AF	EF	IF	DF	GF	HF	0.2
BI	CI	FI	AI	EI	II	DI	GI	HI	0.3
BB	CB	FB	AB	EB	IB	DB	GB	HB	0.4
BE	CE	FE	AE	EE	IE	DE	GE	HE	0.5
BH	CH	FH	AH	EH	IH	DH	GH	HH	0.6
BA	CA	FA*	AA	EA	IA	DA	GA	HA	0.7
BD	CD	FD	AD	ED	ID	DD	GD	HD	0.8
BG	CG	FG	AG	EG	IG	DG	GG	HG	0.9

*白露降。 White dew falls.

Comment: Autumn and the light hoar frost suggest Withdrawal. There is profit in disengaging, retreating. To realize this seemingly contradictory truth, one must turn inward and examine one's position with sincerity. There is no profit in going anywhere. If one should be demoted at this time, it is through no personal fault and will turn out to be for the best. Do not take it to heart.

翕 六之八

翕。利徵師。會同，吉。財聚民散；財散民聚。

Agreement[84] *Xi* 6.8

Agreement. Profit in calling out the army. Gather allies—Good Luck. When wealth is accumulated, people disperse; where wealth is dispersed, people assemble.[85]

數曰:

The Numbers say:

9	8	7	6	5	4	3	2	1	
BF	CF	FF	AF	EF	IF	DF	GF	HF	0.1
BI	CI	FI	AI	EI	II	DI	GI	HI	0.2
BH	CH	FH	AH	EH	IH	DH	GH	HH	0.3
BC	CC	FC	AC	EC	IC	DC	GC	HC	0.4
BE	CE	FE	AE	EE	IE	DE	GE	HE	0.5
BG	CG	FG	AG	EG	IG	DG	GG	HG	0.6
BB	CB	FB	AB	EB	IB	DB	GB	HB	0.7
BA	CA*	FA	AA	EA	IA	DA	GA	HA	0.8
BD	CD	FD	AD	ED	ID	DD	GD	HD	0.9

*寒蟬鳴。 Cold cicada chirrups.

Comment: There is Agreement among your people. This is a time for large scale aggressive action. There is advantage to be gained by mustering the troops. Gathering allies from outside your domain brings Good Luck. Remember that when the spoils are kept for the few, many will abandon the cause; where the wealth is shared, the people will rally around.

遠 六之九

遠。利有攸往。不于其身[R]，于其子孫[R]；不于其家，于其國人[R]。

Distance *Yuan* 6.9

Distance. Profit in going somewhere. If not pertaining to oneself, pertaining to one's child, grandchild. If not pertaining to one's family, pertaining to persons of one's country.

數曰:

The Numbers say:

9	8	7	6	5	4	3	2	1	
BI	CI	FI	AI	EI	II	DI	GI	HI	0.1
BH	CH	FH	AH	EH	IH	DH	GH	HH	0.2
BG	CG	FG	AG	EG	IG	DG	GG	HG	0.3
BF	CF	FF	AF	EF	IF	DF	GF	HF	0.4
BE	CE	FE	AE	EE	IE	DE	GE	HE	0.5
BD	CD	FD	AD	ED	ID	DD	GD	HD	0.6
BC	CC	FC	AC	EC	IC	DC	GC	HC	0.7
BB	CB	FB	AB	EB	IB	DB	GB	HB	0.8
BA	CA	FA	AA	EA	IA	DA	GA	HA	0.9

Comment: Travel at a Distance—there is profit in going somewhere. If the purpose of travel is not for one's personal benefit, it will be for the benefit of one's children and grandchildren. Or, if it is not for the benefit of one's own family, it will be for the benefit of one's countrymen.

迅 七之一

迅。吉。雷風之歘，震撓萬物。君子威德，神化不測。

Quickness *Xun* 7.1

Quickness. Good Luck. Rush of thunder and wind shake, agitate the myriad things. The awesome virtue of the noble-minded person enacts miraculous transformations immeasurable.

數曰:

The Numbers say:

9	8	7	6	5	4	3	2	1	
GA	DA	AA	HA	EA	BA	IA	FA	CA*	0.1
GB	DB	AB	HB	EB	BB	IB	FB	CB	0.2
GC	DC	AC	HC	EC	BC	IC	FC	CC	0.3
GD	DD	AD	HD	ED	BD	ID	FD	CD	0.4
GE	DE	AE	HE	EE	BE	IE	FE	CE	0.5
GF	DF	AF	HF	EF	BF	IF	FF	CF	0.6
GG	DG	AG	HG	EG	BG	IG	FG	CG	0.7
GH	DH	AH	HH	EH	BH	IH	FH	CH	0.8
GI	DI	AI	HI	EI	BI	II	FI	CI	0.9

*處暑；鷹乃祭鳥。 In the Heat;[86] hawks now sacrifice other birds.

Comment: In Quickness there is Good Luck. A rush of thunder and wind rattles and rouses the myriad things. Likewise, the force of the noble-minded person's awe-inspiring virtue can in an instant effect seemingly miraculous changes that are beyond comprehension.

懼 七之二

懼。有孚。惕，厲；終吉。君子畏命；小人畏令。酒食讌樂，凶。

Anxiety *Ju* 7.2

Anxiety. Have faith. Worries; threat—in the end, Good Luck. Noble-minded person is in awe of fate.[87] Small-minded person is awed by a command. Wine, food, feasts, pleasures—Bad Luck.

數曰:

The Numbers say:

9	8	7	6	5	4	3	2	1	
GD	DD	AD	HD	ED	BD	ID	FD	CD	0.1
GA	DA	AA	HA	EA	BA	IA	FA*	CA	0.2
GB	DB	AB	HB	EB	BB	IB	FB	CB	0.3
GG	DG	AG	HG	EG	BG	IG	FG	CG	0.4
GE	DE	AE	HE	EE	BE	IE	FE	CE	0.5
GC	DC	AC	HC	EC	BC	IC	FC	CC	0.6
GH	DH	AH	HH	EH	BH	IH	FH	CH	0.7
GI	DI	AI	HI	EI	BI	II	FI	CI	0.8
GF	DF	AF	HF	EF	BF	IF	FF	CF	0.9

*天地始肅 Heaven and earth begin the withering.

Comment: One notes with Anxiety that plants show signs of withering. One must keep faith in processes of nature, wherein decay is always balanced by the regeneration that ever follows. One worries about the threat of decline, but in the end there is Good Luck. The noble-minded person always holds in awe the role of fate in ordaining outcomes. The small-minded person is only in awe of his superior's orders. Attempting to counter anxiety with food and wine, feasting and indulgence in pleasures—these bring Bad Luck.

除 七之三

除。稊稗，既去，嘉穀斯登。不利作興。君子攸行。

Ridding *Chu* 7.3

Ridding. Cast out the weeds, the wild grass, and excellent grain will rise there. No profit in lifting up. Noble-minded person takes action.

數曰:

The Numbers say:

9	8	7	6	5	4	3	2	1	
GG	DG	AG	HG	EG	BG	IG	FG	CG	0.1
GD	DD	AD	HD	ED	BD	ID	FD	CD	0.2
GA	DA	AA	HA	EA	BA	IA*	FA	CA	0.3
GH	DH	AH	HH	EH	BH	IH	FH	CH	0.4
GE	DE	AE	HE	EE	BE	IE	FE	CE	0.5
GB	DB	AB	HB	EB	BB	IB	FB	CB	0.6
GI	DI	AI	HI	EI	BI	II	FI	CI	0.7
GF	DF	AF	HF	EF	BF	IF	FF	CF	0.8
GC	DC	AC	HC	EC	BC	IC	FC	CC	0.9

*禾乃登 Grain is now tall.

Comment: The grain is tall, and is time for Ridding the fields of choking weeds and wild grasses—cast them out, and healthy edible crops can grow in their place. No profit in rising to prominence. Still, the noble-minded person takes action.

弱 七之四

弱。丈人，厲；小子，吉。不附不植；附則附失。艱正，無咎。

Weakness *Ruo* 7.4

Weakness. Mature person—Threat. Small child—Good Luck. Do not attach to what has not been planted; attach, then attach to loss. Difficult propriety—no Error.

數曰:

The Numbers say:

9	8	7	6	5	4	3	2	1	
GB	DB	AB	HB	EB	BB	IB	FB	CB	0.1
GC	DC	AC	HC	EC	BC	IC	FC	CC	0.2
GF	DF	AF	HF	EF	BF	IF	FF	CF	0.3
GA	DA	AA	HA	EA	BA*	IA	FA	CA	0.4
GE	DE	AE	HE	EE	BE	IE	FE	CE	0.5
GI	DI	AI	HI	EI	BI	II	FI	CI	0.6
GD	DD	AD	HD	ED	BD	ID	FD	CD	0.7
GG	DG	AG	HG	EG	BG	IG	FG	CG	0.8
GH	DH	AH	HH	EH	BH	IH	FH	CH	0.9

*白露；鴻雁來。 White Dew;[88] swans and geese come.[89]

Comment: Weakness sets in as autumn progresses. The mature person, one who has seen many autumns and the winters that followed, understands the threat. To the child, the change of seasons brings new fruits daily and new games; that they don't worry about the future is their Good Luck. Do not become attached to what has not been planted—it is too late for that. Were you to be so attached, you would be attached to things that are sure to be lost. In propriety that is difficult to put into action, there is no error.

疾 七之五

疾。節飲食；謹起居；無攸害。

Sickness *Ji* 7.5

Sickness. Limit eating, drinking;[90] take caution in leaving your dwelling—have no harm.

數曰:

The Numbers say:

9	8	7	6	5	4	3	2	1	
GI	DF	AC	HH	EE	BB	IG	FD	CA	0.1
GH	DI	AF	HG	EE	BC	ID	FA	CB	0.2
GG	DH	AI	HD	EE	BF	IA	FB	CC	0.3
GF	DC	AB	HI	EE	BA	IH	FG	CD	0.4
GA	DA	AA	HA	EA*	BA	IA	FA	CA	0.5
GD	DG	AH	HA	EE	BI	IB	FC	CF	0.6
GC	DB	AA	HF	EE	BD	II	FH	CG	0.7
GB	DA	AD	HC	EE	BG	IF	FI	CH	0.8
GA	DD	AG	HB	EE	BC	IC	FF	CI	0.9

*玄鳥歸。 Swallows return.

Comment: After weakness comes Sickness. By taking care to regulate what one eats and drinks and being cautious in leaving the safety of one's home, one will come to no harm.

競 七之六

競。烏走兎從[R]。麥生茸茸[R]。老夫丰容[R]：争訟逆，凶[R]。

Competing *Jing* 7.6

Competing. Crow walks, rabbit follows. Wheat sprouts soft, soft. Old husband's handsome face: struggle, dispute, opposing—Bad Luck.

數曰:

The Numbers say:

9	8	7	6	5	4	3	2	1	
GH	DH	AH	HH	EH	BH	IH	FH	CH	0.1
GG	DG	AG	HG	EG	BG	IG	FG	CG	0.2
GD	DD	AD	HD	ED	BD	ID	FD	CD	0.3
GI	DI	AI	HI	EI	BI	II	FI	CI	0.4
GE	DE	AE	HE	EE	BE	IE	FE	CE	0.5
GA	DA	AA	HA*	EA	BA	IA	FA	CA	0.6
GF	DF	AF	HF	EF	BF	IF	FF	CF	0.7
GC	DC	AC	HC	EC	BC	IC	FC	CC	0.8
GB	DB	AB	HB	EB	BB	IB	FB	CB	0.9

*群鳥養羞。 Flocks of birds store up sustenance.

Comment: When the grain is ripe and falling, animals begin Competing for it—where the crow walks in the fields, the rabbit will also follow. The wheat is ripe with soft sprouts wisping from the ears. If an old man's handsome and flourishing face leads him to mischief with women, it will end in marital strife, and that is Bad Luck.

分 七之七

分。長短均平[R]；潮駛月盈[R]。君子利正[R]；小人勿乘[R]。

Dividing *Fen* 7.7

Dividing. Long, short in equal balance. Tides roll; moon is full. Noble-minded person profits from propriety; small-minded person cannot take advantage of it.

數曰:

The Numbers say:

9	8	7	6	5	4	3	2	1	
GC	DC	AC	HC	EC	BC	IC	FC	CC	0.1
GF	DF	AF	HF	EF	BF	IF	FF	CF	0.2
GI	DI	AI	HI	EI	BI	II	FI	CI	0.3
GB	DB	AB	HB	EB	BB	IB	FB	CB	0.4
GE	DE	AE	HE	EE	BE	IE	FE	CE	0.5
GH	DH	AH	HH	EH	BH	IH	FH	CH	0.6
GA	DA	AA*	HA	EA	BA	IA	FA	CA	0.7
GD	DD	AD	HD	ED	BD	ID	FD	CD	0.8
GG	DG	AG	HG	EG	BG	IG	FG	CG	0.9

*秋分；雷始收聲。 Autumn Equinox;[91] thunder first gathers in its sound.

Comment: The autumn equinox arrives, Dividing day and night back into two equal parts. When the moon is full, the tides surge strongest. The noble-minded person finds profit in the propriety that arises from measuring these natural phenomena. The small-minded person cannot take advantage of this kind of knowledge.

訟 七十八

訟。內訟，吉。勿有言。不利有攸往。

Dispute *Song* 7.8

Dispute. Internal dispute—Good Luck. Have no words. No profit going anywhere.

數曰:

The Numbers say:

9	8	7	6	5	4	3	2	1	
GF	DF	AF	HF	EF	BF	IF	FF	CF	0.1
GI	DI	AI	HI	EI	BI	II	FI	CI	0.2
GH	DH	AH	HH	EH	BH	IH	FH	CH	0.3
GC	DC	AC	HC	EC	BC	IC	FC	CC	0.4
GE	DE	AE	HE	EE	BE	IE	FE	CE	0.5
GG	DG	AG	HG	EG	BG	IG	FG	CG	0.6
GB	DB	AB	HB	EB	BB	IB	FB	CB	0.7
GA	DA*	AA	HA	EA	BA	IA	FA	CA	0.8
GD	DD	AD	HD	ED	BD	ID	FD	CD	0.9

*蟄蟲坯戶。 Hibernating insects tamp their doors.

Comment: Dividing is followed by Dispute. If dispute is inside one's household, that is Good Luck. One can end it by keeping silent. But there is no profit in going somewhere outside, where the chance of conflict escalates.

收 七之九

收。一氣酉揫[R]；百物歛收[R]。君子反身，放心是求[R]。歛藏吉。

Harvest *Shou* 7.9

Harvest. Elemental unity approaches the gathering in. The hundred things are gleaned, harvested. Noble-minded person examines self and thereby seeks the wandering mind.[92] Gleaning, storing–Good Luck.

數曰:

The Numbers say:

9	8	7	6	5	4	3	2	1	
GI	DI	AI	HI	EI	BI	II	FI	CI	0.1
GH	DH	AH	HH	EH	BH	IH	FH	CH	0.2
GG	DG	AG	HG	EG	BG	IG	FG	CG	0.3
GF	DF	AF	HF	EF	BF	IF	FF	CF	0.4
GE	DE	AE	HE	EE	BE	IE	FE	CE	0.5
GD	DD	AD	HD	ED	BD	ID	FD	CD	0.6
GC	DC	AC	HC	EC	BC	IC	FC	CC	0.7
GB	DB	AB	HB	EB	BB	IB	FB	CB	0.8
GA	DA	AA	HA	EA	BA	IA	FA	CA	0.9

Comment: Harvest time. The cyclical course of the single elemental *Qi* comes to the season of gathering in, when one picks and harvests the many kinds of growing crops. The noble-minded person now examines himself and seeks to concentrate the wandering mind. Glean the fields and store up in the granaries—that brings Good Luck.

實 八之一

實。碩果于叢[R]。仁復于宮[R]；應感不窮[R]。永正，吉。

Fruits *Shi* 8.1

Fruits. Best fruit[93] grows in bunches. Benevolence returns to the palace; it responds to what inspires it unceasingly. Perpetual propriety—Good Luck.

數曰:

The Numbers say:

9	8	7	6	5	4	3	2	1	
DA	AA	BA	GA	EA	CA	HA	IA	FA*	0.1
DB	AB	BB	GB	EB	CB	HB	IB	FB	0.2
DC	AC	BC	GC	EC	CC	HC	IC	FC	0.3
DD	AD	BD	GD	ED	CD	HD	ID	FD	0.4
DE	AE	BE	GE	EE	CE	HE	IE	FE	0.5
DF	AF	BF	GF	EF	CF	HF	IF	FF	0.6
DG	AG	BG	GG	EG	CG	HG	IG	FG	0.7
DH	AH	BH	GH	EH	CH	HH	IH	FH	0.8
DI	AI	BI	GI	EI	CI	HI	II	FI	0.9

*水始涸。 Waters begin to dry up.

Comment: Fruits hang ripe on tree and vine; the best fruit grows in clusters and bunches. Once more as of old benevolence reigns in the palace, responding to the supplications that inspire it without end. In perpetual propriety there is Good Luck.

賔 八之二

賔。俊民用章[R]；觀國之光[R]。利賔于王[R]；大有吉。慶。

Sojourn *Bin* 8.2

Sojourn. The talented among the people are distinguished;[94] observe the country's brilliance. Profit in sojourning with the king;[95] Have great Good Luck. Congratulations.

數曰:

The Numbers say:

9	8	7	6	5	4	3	2	1	
DD	AD	BD	GD	ED	CD	HD	ID	FD	0.1
DA	AA	BA	GA	EA	CA	HA	IA*	FA	0.2
DB	AB	BB	GB	EB	CB	HB	IB	FB	0.3
DG	AG	BG	GG	EG	CG	HG	IG	FG	0.4
DE	AE	BE	GE	EE	CE	HE	IE	FE	0.5
DC	AC	BC	GC	EC	CC	HC	IC	FC	0.6
DH	AH	BH	GH	EH	CH	HH	IH	FH	0.7
DI	AI	BI	GI	EI	CI	HI	II	FI	0.8
DF	AF	BF	GF	EF	CF	HF	IF	FF	0.9

*寒露；鴻雁來賓。 Cold Dew;[96] swans and geese come to sojourn.[97]

Comment: After the harvest, one Sojourns at the estates of the powerful. Now are those of talent accorded distinction by their invitations; now can one see the country's brilliant prowess. There is special profit for those who sojourn with the king—they have great Good Luck. They are to be congratulated.

危 八之三

危。厲。無咎。知險而懼懼。不失正；自天有命。不利涉大川。

Peril *Wei* 8.3

Peril. Threat. No Error. Know adversity and be anxious, anxious. Do not lose propriety—from heaven comes fate's command. No profit in fording the great river.

數曰:

The Numbers say:

9	8	7	6	5	4	3	2	1	
DG	AG	BG	GG	EG	CG	HG	IG	FG	0.1
DD	AD	BD	GD	ED	CD	HD	ID	FD	0.2
DA	AA	BA	GA	EA	CA	HA*	IA	FA	0.3
DH	AH	BH	GH	EH	CH	HH	IH	FH	0.4
DE	AE	BE	GE	EE	CE	HE	IE	FE	0.5
DB	AB	BB	GB	EB	CB	HB	IB	FB	0.6
DI	AI	BI	GI	EI	CI	HI	II	FI	0.7
DF	AF	BF	GF	EF	CF	HF	IF	FF	0.8
DC	AC	BC	GC	EC	CC	HC	IC	FC	0.9

*雀入大水為蛤。 Sparrows enter the great waters and become clams.

Comment: There is Peril. There are threats. There will be no error. Still, one must recognize the time is adverse and remain alert and on guard. Do not waver from propriety; know that whatever occurs is the destiny ordained by heaven. There is nothing to be gained by taking a risk like crossing the great river.

堅 八之四

堅。利有攸往。剛健篤實：義之所出；物莫能屈。攻城陷陣，凶。

Firmness *Jian* 8.4

Firmness. Profit in going somewhere. Hardness, aggressiveness, seriousness, solidity:[98] from these emerge morality; they are things nothing can bend. Assail wall, decimate troops—Bad Luck.

數曰:

The Numbers say:

9	8	7	6	5	4	3	2	1	
DB	AB	BB	GB	EB	CB	HB	IB	FB	0.1
DC	AC	BC	GC	EC	CC	HC	IC	FC	0.2
DF	AF	BF	GF	EF	CF	HF	IF	FF	0.3
DA	AA	BA	GA	EA	CA*	HA	IA	FA	0.4
DE	AE	BE	GE	EE	CE	HE	IE	FE	0.5
DI	AI	BI	GI	EI	CI	HI	II	FI	0.6
DD	AD	BD	GD	ED	CD	HD	ID	FD	0.7
DG	AG	BG	GG	EG	CG	HG	IG	FG	0.8
DH	AH	BH	GH	EH	CH	HH	IH	FH	0.9

*菊有黃華。 Chrysanthemums have yellow flowers.

Comment: Autumn advances, Firmness sets in. There is profit in venturing forth. The time of firmness calls for hardness, aggressiveness, seriousness, solidity—traits that are inflexible and give rise to moral correctness. But if you go to the extent of besieging an enemy city and thereby decimating your forces, it will be Bad Luck.

革 八之五

革。利正。從而革；通不塞。應時而亨[R]；金道乃行[R]。疾病，凶。

Change Ge 8.5

Change. Profit in propriety. Follow and change; penetrate, no blockage. Respond to the time and there is Success; act in the Way of metal.[99] Sickness, disease—Bad Luck.

數曰:

The Numbers say:

9	8	7	6	5	4	3	2	1	
DA	AF	BC	GH	EE	CB	HG	ID	FA	0.1
DH	AI	BF	GG	EE	CC	HD	IA	FB	0.2
DG	AH	BI	GD	EE	CF	HA	IB	FC	0.3
DF	AC	BB	GI	EE	CA	HH	IG	FD	0.4
DA	AA	BA	GA	EA*	CA	HA	IA	FA	0.5
DD	AG	BH	GA	EE	CI	HB	IC	FF	0.6
DC	AB	BA	GF	EE	CD	HI	IH	FG	0.7
DB	AA	BD	GC	EE	CG	HF	II	FH	0.8
DA	AD	BG	GB	EE	CH	HC	IF	FI	0.9

*霜降；豺祭獸 Frost Descends;[100] wild dogs sacrifice beasts.

Comment: With the killing frost comes the growth cycle's great Change. There is profit for those with propriety—if one follows the changes, one can still pass through where the way is not blocked. Responding to the conditions posed by this moment in time brings success. This is acting in the Way of metal, the archetype of autumn. Metal changes from solid to liquid in exact compliance with the heat applied to it; it can be cast or hammered into whatever shape that is imposed on it. Sickness and disease at this time is Bad Luck indeed.

報 八之六

報。祭祀，吉。事不宜先宜後。君子有慶。

Repayment *Bao* 8.6

Repayment. Sacrifice, worship—Good Luck. Such service does not benefit what came before; it benefits what comes after. Noble-minded person receives congratulations.

數曰:

The Numbers say:

9	8	7	6	5	4	3	2	1	
DH	AH	BH	GH	EH	CH	HH	IH	FH	0.1
DG	AG	BG	GG	EG	CG	HG	IG	FG	0.2
DD	AD	BD	GD	ED	CD	HD	ID	FD	0.3
DI	AI	BI	GI	EI	CI	HI	II	FI	0.4
DE	AE	BE	GE	EE	CE	HE	IE	FE	0.5
DA	AA	BA	GA*	EA	CA	HA	IA	FA	0.6
DF	AF	BF	GF	EF	CF	HF	IF	FF	0.7
DC	AC	BC	GC	EC	CC	HC	IC	FC	0.8
DB	AB	BB	GB	EB	CB	HB	IB	FB	0.9

*草木黄落。 Grasses and trees yellow and fall.

Comment: The leaves fall and the time of Repayment arrives. One offers back to the powers of nature some of what one has received through sacrifice of part of the harvest and in worship. Thereby one will one have Good Luck. Such offerings have nothing to do with the past; they look ahead to the future, when the annual round will repeat and the seasonal rituals remind one that heaven and earth provide enough and to spare. For this, the noble-minded person will be congratulated.

止 八之七

止。父慈，子孝；兄友，弟恭。思出位，越常，凶。征吝。

Stopping *Zhi* 8.7

Stopping. Father is benevolent; child is filial. Elder sibling is friendly; the younger, respectful. To think of stepping out of these roles, of overstepping what is constant—Bad Luck. Attack—Remorse.

數曰:

The Numbers say:

9	8	7	6	5	4	3	2	1	
DC	AC	BC	GC	EC	CC	HC	IC	FC	0.1
DF	AF	BF	GF	EF	CF	HF	IF	FF	0.2
DI	AI	BI	GI	EI	CI	HI	II	FI	0.3
DB	AB	BB	GB	EB	CB	HB	IB	FB	0.4
DE	AE	BE	GE	EE	CE	HE	IE	FE	0.5
DH	AH	BH	GH	EH	CH	HH	IH	FH	0.6
DA	AA	BA*	GA	EA	CA	HA	IA	FA	0.7
DD	AD	BD	GD	ED	CD	HD	ID	FD	0.8
DG	AG	BG	GG	EG	CG	HG	IG	FG	0.9

*蟄蟲咸俯。 Hibernating insects all turn face down.

Comment: The time of Stopping is the cusp of autumn and winter. In family relationships, the proper attitudes are displayed: fathers treat their children with benevolence and children are filial; elder siblings are friendly to their youngers, who are respectful in their turn. Any idea of not fulfilling these roles, which would be transgressing the established norms of human society, brings Bad Luck. To attack an enemy at this time brings failure and remorse.

戎 八之八

戎。正，吉。戰血元黃[R]；陽亢有傷[R]。君子克臧[R]；惟知之藏[R]。利征伐。

Weapons *Rong* 8.8

Weapons. Propriety—Good Luck. Blood of war is black and yellow.[101] *Yang* overreaches,[102] receives hurt. If the noble-minded person attacks the good,[103] all that will be known is burying. Profit in attacking, striking.

數曰:

The Numbers say:

9	8	7	6	5	4	3	2	1	
DF	AF	BF	GF	EF	CF	HF	IF	FF	0.1
DI	AI	BI	GI	EI	CI	HI	II	FI	0.2
DH	AH	BH	GH	EH	CH	HH	IH	FH	0.3
DC	AC	BC	GC	EC	CC	HC	IC	FC	0.4
DE	AE	BE	GE	EE	CE	HE	IE	FE	0.5
DG	AG	BG	GG	EG	CG	HG	IG	FG	0.6
DB	AB	BB	GB	EB	CB	HB	IB	FB	0.7
DA	AA*	BA	GA	EA	CA	HA	IA	FA	0.8
DD	AD	BD	GD	ED	CD	HD	ID	FD	0.9

*立冬；水始冰。 Advent of Winter;[104] water first freezes.

Comment: Weapons are the harbinger of winter; war comes after the harvest is in. Maintain propriety and there will be Good Luck. The blood of battle blackens the yellow earth; the aggressive *yang* force exceeds its capabilities and is severely wounded. If the noble-minded person were to take the field against the greater good, the outcome would be burying the dead. But there is gain to be gotten in attacking, in striking a blow where it is justified.

結 八之九

結。百穀其成R；庶續其凝R。履霜堅冰R。婚媾，吉。爭訟，凶。

Knotting *Jie* 8.9

Knotting. Hundred grains have come to their completion; all things continue on to their congealing. Tread on frost, firm ice.[105] Marrying—Good Luck. Struggle, dispute—Bad Luck.

數曰:

The Numbers say:

9	8	7	6	5	4	3	2	1	
DI	AI	BI	GI	EI	CI	HI	II	FI	0.1
DH	AH	BH	GH	EH	CH	HH	IH	FH	0.2
DG	AG	BG	GG	EG	CG	HG	IG	FG	0.3
DF	AF	BF	GF	EF	CF	HF	IF	FF	0.4
DE	AE	BE	GE	EE	CE	HE	IE	FE	0.5
DD	AD	BD	GD	ED	CD	HD	ID	FD	0.6
DC	AC	BC	GC	EC	CC	HC	IC	FC	0.7
DB	AB	BB	GB	EB	CB	HB	IB	FB	0.8
DA	AA	BA	GA	EA	CA	HA	IA	FA	0.9

Comment: Knotting is the tying up of ends. Now the various grains have completed their growth cycle and hardened into seeds; all of nature is drawing up into its dormant state. One treads on hard frost and firm ice. In marrying at this time—tying oneself to another—there is Good Luck; in struggle and dispute there is Bad Luck.

養 九之一

養。惟心，亨。求口實。大人，大體；小人，小體。

Nourishing *Yang* 9.1

Nourishing. Contemplate the mind—Success.[106] Seek food for the mouth.[107] Great person embodies greatness; small-minded person embodies smallness.

數曰:

The Numbers say:

9	8	7	6	5	4	3	2	1	
AA	BA	CA	DA	EA	FA	GA	HA	IA*	0.1
AB	BB	CB	DB	EB	FB	GB	HB	IB	0.2
AC	BC	CC	DC	EC	FC	GC	HC	IC	0.3
AD	BD	CD	DD	ED	FD	GD	HD	ID	0.4
AE	BE	CE	DE	EE	FE	GE	HE	IE	0.5
AF	BF	CF	DF	EF	FF	GF	HF	IF	0.6
AG	BG	CG	DG	EG	FG	GG	HG	IG	0.7
AH	BH	CH	DH	EH	FH	GH	HH	IH	0.8
AI	BI	CI	DI	EI	FI	GI	HI	II	0.9

*地始凍。 Ground first freezes.

Comment: In winter Nourishing proceeds from what has been stored up. Contemplate what goes on in your own mind, and you will have Success. One important thing the mind does is to seek the foods that go in the mouth and nourish the body. The greatness of the great person shows through in his impressive bearing; the smallness of the small-minded person shows through in his paltry bearing.

遇 九之二

遇。吉。非龍、非彲、非虎、非羆為周之師。自天祐之。勿娶女，[108]凶。

Meeting *Yu* 9.2

Meeting. Good Luck. It is not dragon, is not basilisk, is not tiger, is not bear that serves the army of Zhou.[109] Protection comes from heaven.[110] Not taking a mate—Bad Luck.

數曰:

The Numbers say:

9	8	7	6	5	4	3	2	1	
AD	BD	CD	DD	ED	FD	GD	HD	ID	0.1
AA	BA	CA	DA	EA	FA	GA	HA*	IA	0.2
AB	BB	CB	DB	EB	FB	GB	HB	IB	0.3
AG	BG	CG	DG	EG	FG	GG	HG	IG	0.4
AE	BE	CE	DE	EE	FE	GE	HE	IE	0.5
AC	BC	CC	DC	EC	FC	GC	HC	IC	0.6
AH	BH	CH	DH	EH	FH	GH	HH	IH	0.7
AI	BI	CI	DI	EI	FI	GI	HI	II	0.8
AF	BF	CF	DF	EF	FF	GF	HF	IF	0.9

*雉入大水為蜃。 Pheasants enter the great waters and become mollusks.

Comment: A chance Meeting occurs and brings Good Luck. When the future King Wen, founder of the Zhou Dynasty, happened upon the strategist who would help win the throne for his family, the meeting came after a divination reading that the king would not get supernatural assistance from a dragon, basilisk, tiger, or bear. Instead, decisive assistance came from a capable person living as a hermit rather than serving the depraved ruler that King Wen would replace. His protection thus came from heaven, and a seeming random event proved to be preordained. Marriage becomes more urgent—not taking a mate at this time is Bad Luck.

勝 九之三

勝。厲。正，吉。利涉大川。君子以知；小人以力。

Triumph *Sheng* 9.3

Triumph. Threat. Propriety—Good Luck. Profit in fording the great river. Noble-minded person uses knowledge; small-minded person uses strength.

數曰:

The Numbers say:

9	8	7	6	5	4	3	2	1	
AG	BG	CG	DG	EG	FG	GG	HG	IG	0.1
AD	BD	CD	DD	ED	FD	GD	HD	ID	0.2
AA	BA	CA	DA	EA	FA	GA*	HA	IA	0.3
AH	BH	CH	DH	EH	FH	GH	HH	IH	0.4
AE	BE	CE	DE	EE	FE	GE	HE	IE	0.5
AB	BB	CB	DB	EB	FB	GB	HB	IB	0.6
AI	BI	CI	DI	EI	FI	GI	HI	II	0.7
AF	BF	CF	DF	EF	FF	GF	HF	IF	0.8
AC	BC	CC	DC	EC	FC	GC	HC	IC	0.9

*小雪；虹藏不見。 Small Snow;[111] rainbow stored up, not seen.

Comment: At this point in the waning year one can feel Triumph. There are threats, but propriety brings Good Luck. There is profit to be had in taking a risk like crossing the great river on one's frontier. The noble-minded person uses wisdom to attain Triumph; the small-minded person relies on brute force.

囚 九之四

囚。厲。利用獄。[112]不利有攸往。

Imprisonment *Qiu* 9.4

Imprisonment. Threat. Profit in bringing lawsuits. No profit in going anywhere.

數曰:

The Numbers say:

9	8	7	6	5	4	3	2	1	
AB	BB	CB	DB	EB	FB	GB	HB	IB	0.1
AC	BC	CC	DC	EC	FC	GC	HC	IC	0.2
AF	BF	CF	DF	EF	FF	GF	HF	IF	0.3
AA	BA	CA	DA	EA	FA*	GA	HA	IA	0.4
AE	BE	CE	DE	EE	FE	GE	HE	IE	0.5
AI	BI	CI	DI	EI	FI	GI	HI	II	0.6
AD	BD	CD	DD	ED	FD	GD	HD	ID	0.7
AG	BG	CG	DG	EG	FG	GG	HG	IG	0.8
AH	BH	CH	DH	EH	FH	GH	HH	IH	0.9

*天氣上騰地氣下降。 Heaven's elements fly aloft; earth's elements descend below.

Comment: The waning year brings Imprisonment when one's aggressive efforts are countered. This is a threat. Now there is profit in seeking justice in court. There is no profit in trying to escape and run away.

壬 九之五

壬。惟水之神：外暗，内明。君子休休；小人包羞。妊娠，吉。

Cunning *Ren*[113] 9.5

Cunning. Consider the spirit of water: outside dark, inside light. Noble-minded person—Felicity, Felicity. Small-minded person is wrapped in shame. Conception, pregnancy—Good Luck.

數曰:

The Numbers say:

9	8	7	6	5	4	3	2	1	
AI	BF	CC	DH	EE	FB	GG	HD	IA	0.1
AH	BI	CF	DG	EE	FC	GD	HA	IB	0.2
AG	BH	CI	DD	EE	FF	GA	HB	IC	0.3
AF	BC	CB	DI	EE	FA	GH	HG	ID	0.4
AA	BA	CA	DA	EA*	FA	GA	HA	IA	0.5
AD	BG	CH	DA	EE	FI	GB	HC	IF	0.6
AC	BB	CA	DF	EE	FD	GI	HH	IG	0.7
AB	BA	CD	DC	EE	FG	GF	HA	IH	0.8
AA	BD	CG	DB	EE	FH	GC	HF	II	0.9

*閉塞而成冬。 Close up the passes and realize winter.

Comment: In the ninth and closing period of the annual round, one is surrounded by treachery, and it pays to be Cunning oneself. One should meditate on the spirit of water, which is the archetype of winter: on the surface it may appear dark, but within it is crystal clear and translucent. For the noble-minded person, slyness brings felicity; the small-minded person gets trapped in attempts at cleverness and ends up wrapped in shame. During this time, being with child is Good Luck.

固 九之六

固。正，静而一R；為物之極R。龍蛇之蟄R不知不識R。吉。

Rigidity *Gu* 9.6

Rigidity. Propriety, stillness, and unity; these are the peak of things. Hibernation of dragon, snake; no knowing, no understanding. Good Luck.

數曰:

The Numbers say:

9	8	7	6	5	4	3	2	1	
AH	BH	CH	DH	EH	FH	GH	HH	IH	0.1
AG	BG	CG	DG	EG	FG	GG	HG	IG	0.2
AD	BD	CD	DD	ED	FD	GD	HD	ID	0.3
AI	BI	CI	DI	EI	FI	GI	HI	II	0.4
AE	BE	CE	DE	EE	FE	GE	HE	IE	0.5
AA	BA	CA	DA*	EA	FA	GA	HA	IA	0.6
AF	BF	CF	DF	EF	FF	GF	HF	IF	0.7
AC	BC	CC	DC	EC	FC	GC	HC	IC	0.8
AB	BB	CB	DB	EB	FB	GB	HB	IB	0.9

*大雪；鶡旦不鳴。 Great Snow;[114] long-tailed pheasant does not call.

Comment: Deep snow blankets the world, which freezes into Rigidity. In propriety, be still and attain unity within—this is the highest point that any being can experience. This state is like that of the dragon or snake when they are rigid in hibernation; one passes beyond the flow of thoughts that constitutes what one calls "knowing" and "understanding." This is Good Luck.

移 九之七

移。功成而退[R]；居亢則悔[R]。利有攸往。守常，凶。

Shifting *Yi* 9.7

Shifting. Having achieved merit, retire; to stay and overreach makes for regret. Profit in going somewhere. Guarding constancy—Bad Luck.

數曰:

The Numbers say:

9	8	7	6	5	4	3	2	1	
AC	BC	CC	DC	EC	FC	GC	HC	IC	0.1
AF	BF	CF	DF	EF	FF	GF	HF	IF	0.2
AI	BI	CI	DI	EI	FI	GI	HI	II	0.3
AB	BB	CB	DB	EB	FB	GB	HB	IB	0.4
AE	BE	CE	DE	EE	FE	GE	HE	IE	0.5
AH	BH	CH	DH	EH	FH	GH	HH	IH	0.6
AA	BA	CA*	DA	EA	FA	GA	HA	IA	0.7
AD	BD	CD	DD	ED	FD	GD	HD	ID	0.8
AG	BG	CG	DG	EG	FG	GG	HG	IG	0.9

*虎始交. Tigers first mate.

Comment: In a time when it seems time has stopped, it is time for Shifting, moving about. On the one hand, one's achievements are complete, and one should draw back from the fray—to stay engaged and risk overreaching makes for regret. On the other, there is profit to be gained from venturing out. To keep to the tried and true approaches, to continue in the same direction is Bad Luck.

墮 九之八

墮。物極於上，必復於下。君子下下，吉。

Falling *Duo* 9.8

Falling. Things that peak above must return below. Noble-minded person below, below—Good Luck.

數曰:

The Numbers say:

9	8	7	6	5	4	3	2	1	
AF	BF	CF	DF	EF	FF	GF	HF	IF	0.1
AI	BI	CI	DI	EI	FI	GI	HI	II	0.2
AH	BH	CH	DH	EH	FH	GH	HH	IH	0.3
AC	BC	CC	DC	EC	FC	GC	HC	IC	0.4
AE	BE	CE	DE	EE	FE	GE	HE	IE	0.5
AG	BG	CG	DG	EG	FG	GG	HG	IG	0.6
AB	BB	CB	DB	EB	FB	GB	HB	IB	0.7
AA	BA*	CA	DA	EA	FA	GA	HA	IA	0.8
AD	BD	CD	DD	ED	FD	GD	HD	ID	0.9

*荔挺出。 *Liting* iris comes out.

Comment: Falling. When things reach their peak on high, they must fall back below. So the noble-minded person hunkers down, stays out of sight below even the seeming lowest point—in this there is Good Luck.

終 九之九

終。吉。茲闔之窮[R]；斯闢之通 [R]。君子令終[R]。

End *Zhong* 9.9

End. Good Luck. The exhausting of the close of this is the opening of the way to that. Noble-minded person comes to a good end.[115]

數曰:

The Numbers say:

9	8	7	6	5	4	3	2	1	
AI	BI	CI	DI	EI	FI	GI	HI	II	0.1
AH	BH	CH	DH	EH	FH	GH	HH	IH	0.2
AG	BG	CG	DG	EG	FG	GG	HG	IG	0.3
AF	BF	CF	DF	EF	FF	GF	HF	IF	0.4
AE	BE	CE	DE	EE	FE	GE	HE	IE	0.5
AD	BD	CD	DD	ED	FD	GD	HD	ID	0.6
AC	BC	CC	DC	EC	FC	GC	HC	IC	0.7
AB	BB	CB	DB	EB	FB	GB	HB	IB	0.8
AA	BA	CA	DA	EA	FA	GA	HA	IA	0.9

Comment: The End. Ending brings Good Luck, because the end of one thing is another's beginning. The noble-minded person looks to the new and is satisfied in a good ending.

B. The Language of Cai Chen's Number Names

In writing his *Master Plan*, Cai Chen joined a line of *Yijing* emulators that begins with Jiao Gong 焦贛 (1st cen. BCE.), who wrote *Forest of Change* (*Yilin* 易林). Each hexagram in the *Yilin* is linked to the agricultural calendar through the mapping in the *Guaqitu* of Meng Xi, and each heads an array of sixty-four poems—generally four lines of four characters each—one for itself and one for each of the other 63 Hexagrams. The poems are taken to be reflective of the temporal location of the head hexagram in the *Guaqitu*; they contain divination information for every hexagram in the array and were accessed through an unspecified divination method.[116]

In the next generation, Yang Xiong devised the *Taixuanjing* with 81 sets of divinatory readings. These were indexed to a set of tetragrams with a third type of line—a tripartite line—in addition to the unbroken and bipartite types found in the *Yijing*. The names of the tetragrams and their readings were again keyed to the agricultural calendar.[117] Other authors of idiosyncratic divination systems include Guan Lang, whose *Dongji zhenjing* employed 27 of Yang Xiong's 81 tetragrams, and Sima Guang, who wrote the *Qianxu*. As did Cai Chen, Sima Guang grouped divination information under pairs of counting rod numerals, though he used a base ten scheme, while Cai's was base nine. Cai made reference to the *Dongji* and *Qianxu* in the preface to his book.[118]

Although these authors did not articulate such an objective, all appear to have aimed at devising imagery with allusive power like that of the *Yijing*'s but in more consistent and intelligible language. Two millennia of inconclusive debate on the meaning of much of the *Yijing*'s diction testifies to its overall abstractness, a quality that springs from its likely collective authorship by generations of diviners, selective editing of what was at first oral tradition, and the usage of homophones when it was committed to writing.[119] Its telegraphic brevity, lacking auxiliary words or punctuation, results for the most part in a lack of specificity that has consistently frustrated systematic treatment.[120]

By contrast, Cai's legends, like the poems in the *Yilin* and the "Headings (*shou*)" in the *Taixuanjing* can be read with relatively little difficulty to one grounded in the traditional literature of the Confucian school. All three of these books are keyed to the agricultural calendar, and their word choices echo that relationship. While the *Yilin* follows the *Yijing*'s 64 hexagram names, Cai, like Yang Xiong, expanded to 81 reference figures and invented his own names. This allowed Cai to rationalize the names according to his interests in the continuum of the annual round and the complementary opposition in his number theory. Cai also used his legends to advance an ethical view of how best to deal with the vagaries of fate that corresponded to the luck of the draw in his divination method. The following discussion looks in more detail at each of these aspects of Cai's Number Names and legends.

C. The Set of Number Names

Cai intended the Number Names to be a universal set in the sense that its members reflect upon each other according to numerical relationships in the structure of the matrix format and cohere as a meaningful temporal continuum. The skeleton of the continuum is clear in the naming of 1.1, 5.5, and 9.9 as "Origin," "Center," and "End," highlighted in light gray in Table 2.3.

Cai's approach to populating the complete continuum running from 1.1 to 9.9 with apt Names was conditioned in part by his desire to show antithesis in wording that matched the tens complement rule informing his matrix suite. 1.1 Origin and 9.9 End at either end of the suite are clearly intended as verbal opposites in keeping with their numerical opposition. Highlighted in dark gray are Number Names on the vertical 5.n and horizontal N.5 axes. These are paired in numerical opposition across and equidistant from 5.5 Center and are also verbal antonyms. Number Names 4.5 and 6.5, for example, are, respectively, Increase (*yi*) and Decrease (*sun*), which are also verbal opposites in the *Yijing* (Hexagrams 42 and 41). Situated as they are in opposed positions on the numerical grid, they indicate intentionality in placement in accord with Cai's tens complement rule in matrix construction. Likewise, 5.2 Decision (*jue*) and 5.8 Doubt (*yi*), 2.5 Constancy (*chang*) and 8.5 Change (*ge*), and 5.1 Abudance (*shu*) and 5.9 Minimizing (*gua*) would be recognized as antonyms or near opposites in idiomatic Chinese. Others, like 1.5 Straightness and 9.5 Cunning, are less immediately apparent, but notes in the *Caishi jiurushu* indicate that they were understood as opposites, though through the application of obscure secondary or tertiary meanings of characters, like *jen* in 9.5 Cunning.[121] Along the same lines, the Number Names in boldface I judge to show opposite meanings in opposite numerical positions—some clearly so (2.6 Softness and 8.4 Firmness; 3.6 Strength and 7.4 Weakness; 2.9 Flowering and 8.1 Fruit), and others arguably so (e.g., 1.8 Waiting, in the sense of hesitating to engage, and 2.9 Meeting). In all, 25 of the 40 possible pairs in Table 2.3 can be taken as opposites.

Table 2.3: Antonyms among Number Names

9	8	7	6	5	4	3	2	1	
Nourishing	Fruits	Quickness	Decoration	Abundance	Nearness	Visibility	Achievement	Origin	0.1
Meeting	Sojourn	Anxiety	Perversity	Decision	Opening	Taking	Peace	Hiding	0.2
Triumph	Peril	Ridding	Emptiness	Contentment	Progress	Following	Arousal	Guarding	0.3
Imprisonment	Firmness	Weakness	Dimness	Ascenscion	Altruism	Mating	Prayer	Extending	0.4
Cunning	Change	Sickness	Decrease	Center	Increase	Nurturing	Constancy	Straightness	0.5
Rigidity	Repayment	Competing	Utility	Subservience	Distinctness	Strength	Softness	Naiveté	0.6
Shifting	Stopping	Dividing	Withdrawal	Excess	Fullness	Uplifting	Tranquility	Repose	0.7
Falling	Weapons	Dispute	Agreement	Doubt	Bestowing	Gladness	Kinship	Waiting	0.8
End	Knotting	Harvest	Distance	Minimizing	Revelry	Soothing	Flowering	Threat	0.9

Cai's word choices also reflect his sense of the series of 81 Number Names as a continuum. In the *Yijing* first order commentary *Xugua*, a narrative built around the hexagram names is presented as a rationalization for the order in which they occur. The *Xugua* views the hexagram sequence as a dynamic series set in motion by H1 *qian* and H2 *kun*, synonymous with Heaven and Earth, the primal father and mother. From this pair are produced the "myriad things," and each of the succeeding hexagrams is treated as a state or condition that "evolves" from its predecessor. For example, "What fills the midst of Heaven and Earth are only the myriad things. Thus, it evolves into *zhun* [H3]. *Zhun* means 'filling.' *Zhun* means the first birth of things. When things are born, they must be naïve; thus it evolves into *meng* [H4]. *Meng* means 'naiveté.'"[122] One can imagine Cai Chen, who was schooled in the traditional interpretation of the *Yijing* commentaries, attempting, as with his other linguistic gestures, to improve on the *Xugua* model, which in most cases seems forced and unconvincing.

Cai left only one passage with contextual clues to his name choices. It divides into four sections, each headed by the first Number Name, Origin:

1. Origin [1.1] is the beginning of the primal element; Peace [2.2] is the beginning of form; Center [5.5] is the essence of control; Utility [6.6] is the materialization of things; End [9.9] is the culmination of happenings.
2. Origin [1.1] is the precursor of benevolence; Utility [6.6] is the essence of morality; Altruism [4.4] is the enclosure of ritual; Weapons [8.8] is the edge of wisdom; Center [5.5] is the completion of trustworthiness.
3. Origin [1.1] comes near to Center [5.5]; Subservience [5.6] is distanced from Center [5.5]. Near means approaching, and distanced means retreating; near means growth, and distanced means dissolution.
4. The beginning in Origin [1.1] comes back to End [9.9]—thereby know the teaching of death and life.[123]

The five Names in section 1) fall along the 1.1–9.9 diagonal and sketch the life cycle that is Cai's underlying template. Here the seasonal analogy is expressed as the physical development of phenomenal being from the coalescing of elemental force, *Qi*, in (1.1) through its first identifiable manifestation (2.2), individuation (5.5), and full materialization (6.6), and then to its culmination (9.9). The second places the five cardinal Confucian virtues (*wuchang* 五常)[124] on the same diagonal, giving these moral qualities equivalence to the developmental phases marked by natural phenomena. Note that the five are presented in their traditional sequence but placed in a different temporal order on the diagonal. The third statement says that even though 1.1 Origin is 40 nodes distant from 5.5 Center, it is in principle closer to the Center than 5.6 Subservience, which is immediately below Center. Cai's rationale appears to be that Center is the pivotal point between the generative power of the seasonal round, and that the growth that has characterized the cycle up to that point turns to dissolution immediately afterward. Thus, the stage is set for the fourth sentence, which says that whatever begins inevitably comes to an end in the cycle of death and life, phrasing that Cai draws directly from the *Xici* commentary.[125] "Death and life (*sisheng*)" reverses the more common binome "life and death (*shengsi*)" and emphasizes the recurring nature of the process.

D. The Agricultural Calendar in the Placement and Diction of Number Names

In developing his number theory, Cai placed the eight essential solar terms—the solstices, equinoxes, and "advent" days—at the cardinal points of his circular graphic. From there, his folding of the circle's nine arcs into a 9 x 9 matrix leaves them on the 1.1–9.9 diagonal. He said that the Winter Solstice is shared by 1.1 and 9.9—that it pivots, as it were, between the two as a cycle ends and restarts. The other pivotal point is at 5.5, the summer solstice. Between 1.1 and 5.5, the year's productive energy accumulates to its maximum and then disperses toward 9.9. In the latter segment, the autumnal equinox (*qiufen* 秋分) occurs at 7.7, Division, which shares the character (*fen* 分)—a case where the Number Name is clearly derived from the calendar. If the Names in general reflect that natural growth and dispersion, one would expect it to be discernible when they are read against the agricultural calendar as in Table 2.4, where Gregorian month/day dates are shown for the 24 solar terms.

Table 2.4 Number Names Opposed Across Seasonal Continuum

12/22	1.1	Origin				9.9	End
	1.2	Hiding				9.8	Falling
	1.3	Guarding				9.7	Shifting
1/6	1.4	Extending			12/7	9.6	Rigidity
	1.5	Straightness				9.5	Cunning
	1.6	Naiveté				9.4	Imprisonment
1/20	1.7	Repose			11/22	9.3	Triumph
	1.8	Waiting				9.2	Meeting
	1.9	Threat				9.1	Nourishing
	2.1	Achievement				8.9	Knotting
2/4	2.2	Peace			11/7	8.8	Weapons
	2.3	Arousal				8.7	Stopping
	2.4	Prayer				8.6	Repayment
2/19	2.5	Constancy			10/23	8.5	Change
	2.6	Softness				8.4	Firmness
	2.7	Tranquility				8.3	Peril
3/6	2.8	Kinship			10/8	8.2	Sojourn
	2.9	Flowering				8.1	Fruits
	3.1	Visibility				7.9	Harvest
	3.2	Taking				7.8	Dispute
3/21	3.3	Following			9/23	7.7	Dividing
	3.4	Mating				7.6	Competing
	3.5	Nurturing				7.5	Sickness
4/5	3.6	Strength			9/8	7.4	Weakness
	3.7	Uplifting				7.3	Ridding
	3.8	Gladness				7.2	Anxiety
	3.9	Soothing			8/23	7.1	Quickness
4/20	4.1	Nearness				6.9	Distance
	4.2	Opening				6.8	Agreement
	4.3	Progress				6.7	Withdrawal
5/6	4.4	Altruism			8/8	6.6	Utility
	4.5	Increase				6.5	Decrease
	4.6	Distinctness				6.4	Dimness
5/21	4.7	Fullness			7/23	6.3	Emptiness
	4.8	Bestowing				6.2	Perversity
	4.9	Revelry				6.1	Decoration
	5.1	Abundance				5.9	Minimizing
6/6	5.2	Decision			7/7	5.8	Doubt
	5.3	Contentment				5.7	Excess
	5.4	Ascenscion	6/21	5.5 Center		5.6	Subservience

The Number Names in the left three columns proceed from December 22, the winter solstice, to *Mangzhong*, June 6, and mark the growth portion of the continuum, the gradual increase in the amount of daily sunlight. The pivotal term Center 5.5 marks June 21, the summer solstice, and is alone at the bottom of the chart—the only Number Name without a complement. The right three columns, read from bottom to top, represent July through December, during which the amount of sunlight decreases day by day. Antithetical name pairing is shown by gray highlights across the table. From the perspective of those pairs, the ones whose numbering is highlighted in columns 2 and 7 have names that are consistent with their placement in the growth or decay segment—3.6 Strength vs. 7.4 Weakness; 5.2 Decision vs. 5.8 Doubt, etc. Of the 40 pairs of opposites, 25 reflect this relationship, becoming more dense toward the center of the continuum. In addition, five terms in the positive growth segment and seven in the negative segment do not appear to be antonyms of their numeric opposite; nonetheless, they have names that reflect their seasonal implications. The only term in the first part of the continuum that might be taken as having a negative connotation is 1.9 Threat, while 9.3 Triumph would seem to be the lone clear positive outlier among the post-5.5 Number Names. Of the others, 2.9 Flowering and 8.1 Fruit are clearly seasonal, the latter following 7.9 Harvest in late September. Arguments might be made against some of these categorizations, particularly as they are translated. However, there is ample evidence that Cai Chen intentionally selected Number Names that reflected the calendric interests woven into his number theory.

Whether he further intended them to suggest a coherent narrative of growth and decline is more difficult to prove from the Names alone. The names his sequence shares with the *Yijing* do not follow in what the *Xugua* commentary presented as a narrative order.[126] It is not unlikely that the placement of Cai's 1.6 Naiveté early in the sequence echoes the *Yijing* placement of its namesake, H4 *Meng*. Cai's second Number Name, 1.2 Hiding (*qian*), appears in the first line of the first *Yijing* hexagram (H1 *Qian*, 9.1), which reads, "Hidden dragon, do not act (*Qianlong buyong*)." The opening two words of Cai's legend for 1.2 Hiding repeat the same "*buyong*" as in H1 *Qian*, 9.1, confirming the *Yijing*'s influence in this strategic choice.

Selections from the 81 legends taken in sequence also outline the cadence of the natural round from Origin to End: "The hundred things are produced in order; the many matters are achieved in order (2.7 Tranquility). Elemental substance, its shapes and appearances—these are gotten from Heaven. (3.2 Taking). Heaven and earth intertwine threads; the myriad things transform, each to its essence. (3.5 Nurturing). Living elements flow through shapes; types of things are all clearly defined (4.7 Fullness). Things that peak above must return below (9.8 Falling)."

Other legends advance the seasonal analogy by incorporating almanac imagery cited in each matrix from the *Yueling*. 1.9 Threat begins "Birds of prey are swift Threat," wording taken from the *Yueling* citation appended to 1.8 Waiting. 2.3 Arousal occurs when the *Yueling* says, "Hibernating insects are first aroused." 4.3 Progress says "silkworms and mulberry—Good Luck" when the *Yueling* says "Hoopoes alight in the mulberry trees." The character Mi, the name of 4.9 Revelry, is the same term translated as "supple" used in the 4.8 *Yueling* citation, "supple grasses die." The Five Phases (*wuxing*) continuum—wood, fire, earth, metal, and water—is another seasonal analog to which reference is made in "the Way of wood" in 2.9 Flowering and "the Way of metal" in 8.5 Change. The character used to name 9.5 Cunning, *Ren*, is one of the 10 "Heavenly Stems" assigned by convention to water, and its Legend begins, "the spirit of water."

Cai also elided vocabulary in neighboring Number Names—a gesture that recalls the gradual change of seasons that the agricultural calendar highlights by establishing the "advent" of the four seasons several weeks before the solstices and equinoxes. For example, the term translated "brilliantly displayed (*wenming*)" is used in Number Names 2.8 and 2.9, "elemental (*qi*)" in 3.1 and 3.2, "taking (*huo*)" in 3.2 and 3.3, etc.

E. Allusions to the *Yijing* and Other Classic Literature

As the notes to the above translation show, there are numerous allusions to specific *Yijing* passages in the legends. References to *Yijing* hexagram and line words, as well as the primary commentaries *Tuan*, *Xiang*, *Wenyan*, and *Xici* appear in 27 legends.[127] Allusions are found in the first and last Number Names, ending with the last word of the Legend for 9.9 End. In addition to the luck terms discussed in the first section, several mantic phrases common to the *Yijing*, particularly those containing "profit (*li* 利)," are used with some frequency. "Profit in going somewhere (*liyouyouwang* 利有攸往)" appears 12 times, for example, six of them in the negative. "Profit [or not] in crossing the great river (*lishedachuan* 利涉大川)" appears 4 times.

The *Yijing*'s divination information is given in 448 subsets of words, 64 that follow the hexagram names (known as *tuanci* 彖辭) and 6 additional phrases—384 in all—known as "lines (*yao* 爻)" following each of those. The 448 subsets have 9 words each on average. Hexagram words tend to give mainly mantic information, while line words present images from the natural world or human affairs as well as a judgement on relative luck. In general, Cai's legends are longer, at 17 characters on average, and the prognoses are more decisive. For example, the hexagram words for *Yijing*, H46 *Sheng* read: "Ascension: Prodigious success. Of use to see the great person. Do not worry. Southern attack—Good Luck."[128] Cai's Legend for 5.4 Ascension (which uses the same character, *sheng*) has 21 characters in contrast to the *Yijing*'s 12. It reads: "Ascension. Rites shed their brilliance; music is performed. The myriad transformations by these are completed. Profit in seeing the great person. Without words, have understanding. Confidently ascend—great Good Luck." Both refer to seeing "the great person" and both portend Good Luck. But Cai fleshes out his legend with more specific images that add up to a fuller metaphor for the fortunate situation.

Again, *Yijing*, H16 *Yu* reads in 6 characters: "Contentment. Profit in installing baron, moving out troops."[129] Cai's Legend 5.3 of the same name has 13 words and reads, "Contentment. Drink, eat; harmony, joy. From the contentment of the noble-minded person comes Good Luck. From the contentment of the small-minded person comes Bad Luck." The *Yijing* example advances military allusions to explain why one should feel contentment, perhaps because one is secure in being well protected. Cai suggests how one may enjoy his contentment but cautions that one must have the magnanimity of the noble-minded person to turn contentment to advantage.

Cai also made direct reference to several other classical sources in his legends. These included all of the Four Books as delineated by Zhu Xi, the *Shujing*, for which Cai's commentary became the standard, the historical *Shiji* and *Hanshi*, and the ritual work *Liji*. Otherwise, his language has the feel of Han Dynasty writing like the *Yilin*, with a preponderance of four-character clauses, occasionally given as rhyming poetic stanzas. Still, though these devices lend an archaic air to the diction, its meaning is for the most part directly intelligible and would communicate to someone steeped in Confucian lore a fairly unambiguous sense of the divination directive.

F. Qualities and Discipline of the Noble-Minded Person

Throughout the collection of legends, Cai weaves the theme of the noble-minded person (*junzi*) as a paragon to be emulated. The term *junzi* appears frequently in the *Yijing*, the *Shijing*, and the Four Books, and it is developed philosophically in records of the sayings of Kongzi and his disciples. Drawing from the *Lunyu*, Cai wrote that in the time of 4.4 Altruism, "Noble-minded person constrains ego; in rites returns to one's beginning," and of Minimizing 5.9, "Noble-minded person minimizes his errors." In the time of Decoration, 6.1, "Noble-minded person is cautious when alone," an allusion to both the *Zhongyong* and *Daxue*. These instances illustrate an aspect of personal discipline with particular reference to a location in the cyclical temporal pattern. Contrasting the behavioral responses of the noble-minded person versus the coarse small-minded person is Cai's favorite device, used 13 times, for shaping this ethical stance. This comparative trope is used in five instances of *Yijing* wording and six times in its first order commentaries. It is also common in the *Lunyu*, where it appears 15 times, one of which Cai cited in Utility 6.6, where "Noble-minded person thinks in terms of morality. Small-minded person thinks in terms of profit."

To Cai the discipline of which the noble-minded person is capable reveals innate "enlightenment (*ming* 明)"—intelligence that shows forth in one's applied understanding of numbers and images. Cai wrote in his preface, "Without enlightenment toward numbers, one lacks the capacity to discuss images; without enlightenment toward images, how can one have the capacity to know numbers? These two—one can have both but cannot lack either."[130] His sequential uses of "enlightenment" and closely related nuances of *ming* in the legends[131] unfolds the development of this quality in the noble-minded person. In 1.6 Naiveté, "inside there is light," but it is limited so early in the growth cycle by the Naiveté of youth, where its emergence cannot be forced without bad luck. Two nodes later at 1.8 Waiting, there is still "no light;" one must have faith and "center in propriety." At 2.9 Flowering, the quality blossoms and "Through propriety one's cultivation radiates." 4.6 Distinctness comes at a point in the cycle where maturation is apparent and "the designs beneath heaven are brilliantly displayed." Later as decline sets in, there is Doubt 5.8; still, the "noble-minded person enlightens; small-minded person uses negativity."

These references all point back to the opening sentence of the *Daxue*: "The Way of the Great Learning is in bringing to light one's bright virtue (*ming mingde*), in renewing the people, and stopping at the greatest good."[132] "Bright virtue" is the inner light from 1.6 Naiveté. Zhu Xi cites Cheng Yi as saying: "[the first] 'bright' [is a verb that] means 'bring to light.' 'Bright virtue' is [an innate quality] people receive from heaven; it is vacuously spiritual and unobscured, and, because it holds complete all their [individual developmental] Patterns, it responds to the myriad events that occur."[133] That is, one's mind has at its heart an innate capacity—its bright virtue—to correctly evaluate and respond to whatever it encounters. But, because of differences in people's constitutions and the distractions caused by desires, it is at times more or less impeded. It never loses its "brightness," though, and "the student must pursue the brightness whenever it arises in order to return to realizing its original state."[134] Thus, the noble-minded person is "able to be brilliant" when under "threat" in 6.2 Perversity. Then, as in 7.1 Quickness, one can use the power unleashed by one's enlightenment to "renew the people:" "The awesome virtue of the noble-minded person enacts miraculous transformations immeasurable."

The *Daxue* reference also informs the pivotal 5.5 Center, the zenith of the annual cycle, which features the "illustrious, illustrious; great light," which "illuminates" the neighboring countries on all sides. Here the noble-minded person "holds to fullness," or, in *Daxue* terms, "stops in the highest good (*zhiyuzhishan* 止於至善)." By contrast, the lesser mortal—the small-minded person—would at the peak tend to grasp for more and thereby spoil the measure of success he had achieved. As the declining phase of the annual round takes hold, at 6.4 Dimness, the "distinctness" (seen clearly in 4.6 Distinctness, the opposite of 6.4) is "murky," and light is "obscure." There is Good Luck then in the "propriety of the secluded person" and not in overt action. Finally, at 9.5 Cunning, light returns to the internal place it held in 1.6 Naiveté, except that the outside world is darkening with the end of the life cycle rather than being wrapped in the naiveté of youth. But even at this time, the noble-minded person has "felicity, felicity," while the "small-minded person is wrapped in shame."

Cai uses the term righteousness (*zheng* 正) in a quarter of his 81 legends.[135] *Zheng* carries the sense that the power to display one's inner virtue derives from being true to the innate understanding of what is correct in a given situation and the will to act in accordance. Righteousness expresses the enlightenment that is the raw material of a noble-minded person and is shaped through continual learning over the course of one's life. Cai's guidance on how this learning can be accomplished in 1.6 Niaveté: "Profit in studying numbers."

From this perspective, the sequence of 81 Number Names is a "study in numbers," offering a way of conceptualizing distinct moments in the flow of time and defining actions or attitudes that can optimize positive and minimize negative outcomes. One asks advice of the "great person," experienced in negotiating the flow of time, in 1.5 Straightness, 3.7 Following, and 5.4 Ascension. Progress 4.3 also takes up this theme, saying one should "draw near to worthies; cast out the perverse." That legend, too, suggests how understanding the implications of the temporal context is applied. One organizes projects according to one's knowledge of the work and is personally involved, "shouting encouragement." Acting contrary to the times, one is reminded, brings Bad Luck; but tending to the seasonal tasks of husbandry—raising silkworms and cultivating the mulberry leaves on which they feed at this moment in spring—brings Good Luck. Likewise, in the time of 4.5 Increase, one fosters allies, hosting and honoring them. But if one is unused and unemployed, it brings on Bad Luck. Again, one keeps one's faith in the time of Anxiety 7.2 and has Good Luck; this is not a time for wine, food, feasting, and joy. Nor is Weakness 7.4 a time to form attachments. It is a time of threat to an adult—only a small child, to whom it is appropriate, has the Good Luck of parental support in Weakness. In Weakness, as in 5.2 Decision, there is "profit in difficult propriety," which perhaps epitomizes Cai Chen's sense of wisdom in action: to actuate oneself as a noble-minded person, one must apply propriety without ceasing and particularly when the times feel adverse.

G. Architecture of the Images

In presenting a complex, multi-layered imagistic structure, the verbal architecture of the Number Names is a complement to the interlocking suite of numerical matrices. The latter follows consistent rules aimed at synchronizing the progress of the double Good Luck term, AA, cell-by-cell through the 81 graphics. Its location amidst the changing alignment of terms marks the identity of each given matrix, as the Number Name and legend expresses in words what "the numbers say," in Cai's phrase. The Numbers cross-reference at a distance through the relationship of each with its opposites, both in the current matrix and across the series to opposites on the other side of the balancing point in 5.5 Center. These relationships create equations that net to "balance"—zero in mathematical terms, equanimity as a discipline—while providing differentiation in local contexts. It is clear from the number of opposite terms Cai employed in antithetical cells that he intended to highlight that correspondence between numbers and words. His replacement strategy makes for continuity through the 81 matrices, and the seasonal characteristics of Cai's images reflect a similar pattern, many being weighted toward youth and growth early on and later turning to rigidity and decline.

Adjustments in the replacement strategy are required to maintain the AA cell's consistent forward movement; they are applied in the 5.n family and the N.5 member of the other eight matrix sets. The idea of adjusting to achieve a numerical objective is also expressed in the legends' philosophy of adaptation to the changing moment, of learning in the naiveté of youth to prepare for the doubts, perversity, and struggle as well as the harvest and triumph of the adult phase. To Cai, the numbers also said that innate intelligence recognized the righteous course of action, and that, shaped by the discipline of constraining the ego, one's "enlightenment" would lead to the optimal outcome in differing situations. This discipline was the "balance" of the center, the *Luoshu*'s number 5 that mediates among the pairs of opposites and the "equals" sign in Cai's number theory.

More directly than the *Yijing*, Cai's legends offer relatively unambiguous divinatory information, and the forecasts in the paired Luck Spectrum terms of his second operation are blunt. Perhaps this frankness reflects his inclination to take a materialistic view of phenomenal existence, to quantify it and, in his words, "thereby know the teaching of death and life."

Conclusion: The Numbers Say ...

The rules of construction Cai Chen enforced in devising his suite of 81 matrices ensure that the individual members of the series are rich in numerical defining qualities and relationships. When Cai wrote " the Numbers say" for each matrix, he indicated that the configuration of luck terms in its 9 x 9 grid informed its Number Name. It is clear from numerous examples cited above that the Number Names and their legends reflect one overarching numerical structure—a circular template subdivided into four (seasons), 24 (solar periods), and 72 (seasonal periods), as well as 81 individual nodes for the Number Names. The Names reflect the seasonality of their points on the circle, and Cai cites the *Yueling* almanac in both the matrices and in his diction to make this link explicit.

It is not as apparent how the internal numerical structure of a matrix might have contributed to image formation in specific legends, but several instances are suggestive of Cai's thinking. In the case of 1.1 Origin, for example, one might say that the pristine ordinal structure of terms A-I in right-hand terms of columns and left-hand terms of rows is equivalent to the purely numerical grid of 81 terms—the origin of all the rest. This numerical grid remains implicit while column replacement changes take place in each successive overlay; a mathematical universe is evolved and expressed as a domain of discourse. Within that universe, Cai went to significant lengths to locate the cell with with the pair AA, the doubled term Good Luck, in each matrix. AA occupies the first cell of 1.1 Origin, where Cai labeled it "Prodigious Good Luck (*yuanji* 元吉)"[136]— and occurs throughout the series in the cell that carries same numerical coordinates as its home matrix. Additional manoeuvers in his construction rules made AA the second most probable double term in a random draw. Thus, the numbers tend to say "Good Luck," and, indeed, the term occurs a third more frequently than its opposite in the legends. And when AA arrives in cell 9.9 of 9.9 End, the cycle is complete and primed to begin again, as the legend of 9.9 End literally states.

The most frequent pairing is EE, which means "Balance.Balance (*ping.ping* 平平)." Cai said of 5, the number of the Luck Schedule term Balance, "the eight numbers flow cyclically and extrude the various types [of things], but they seek the central 5 and are thus balanced."[137] That Balance was actually the most desirable state was implied in the central cell of the series, matrix 5.5's cell 5.5. There, one would expect to find the E term: 64 matrices have EE in that cell, the 8 N.5 matrices have EA, and the 8 other 5.n matrices have AE. However, the replacement rules forced AA to appear in matrix 5.5, cell 5.5, even though it is the only cell in its matrix 5.5 that has no E term. The character *ping* itself occurs in only two Number Name legends, 2.7 Tranquility and 7.7 Dividing, but its philosophical implication of carefully weighed behavior is a consistent imagistic theme throughout the suite. It is the essential quality of the noble-minded person, regularly displayed in contrast to the intemperance of the small-minded person.

Like 1.1 Origin and 9.9 End, the Name of 5.5 Center matches its position in the numerical sequence. Equivalent to the Luck term Balance in number, 5.5 Center carries several layers of allusive power. In particular, it is surely the "supreme pivot (*huangji*)" that Cai incorporates into the title of his book. At *huangji's locus classicus* in the *Hongfan* chapter of the *Shujing*, Cai's gloss reads, "'Supreme' means what the ruler establishes in place. 'Pivot' is like the 'pivot' in [the heaven's north star or] 'northern pivot.' It has the meaning of "ultimate pivot [point]; it is the name for the standard [against which things are measured]. When the center is in place, what the four directions choose is righteousness."[138] In *Hongfan huangji neipian*, he wrote: "When the pivot is established, the great foundation is in place. When the pivot is understood, then the great use is manifest. With it one can regulate the family and in the family nothing goes unregulated. With it one can govern the country, and in the country nothing is ungoverned. With it one can bring balance to all under heaven, and under heaven nothing is unbalanced."[139]

These words are echoed in Cai's Legend for Number Name 5.5, which reads "Center. Elegant, elegant; great light. Illuminate the four neighbors. Noble-minded person holds to fullness. Small-minded person destroys achievements."[140] Cai made frequent references in the legends to *Zhongyong*, a book that Zhu Xi elevated to prominence among the classics of NeoConfucianism. In defining the title as "The Constant Center," Zhu Xi said, "Center means neither biased nor partial; it is a word that means no error is not corrected. Constant means in balance continually."[141] "Center" and "Balance" are both terms Cai associated with the *Luoshu*'s central number 5, and there can be little doubt that Cai was mindful of the textual referents of both these terms in locating them as the central concepts of his system.[142]

The extent to which Cai's visualization of numerical structures was further incorporated into the architecture of his verbal images is suggested in several other ways in the foregoing discussion. It is not obvious that every word of every legend can be traced back to an origin in its matrix, but in general the segmentation of the circular template may be said to take form in Cai's linguistic choices. Whether or not one grants Cai's contention that number informs his images, however, one must grant that his number theory was a radical departure from the prosaic forays into the construction of magic squares recorded by his contemporary Yang Hui. Indeed, Cai's complex suite of matrices—an extended equation encompassing 81 subequations, each of which is a rule-based semi-magic square—is like nothing imagined before modern times. Devoting himself to "studying the numbers," as he counseled in 1.6 Naiveté as a course of education, he translated the preverbal meaning of his Numbers into concrete Names that made them intelligible. Together, these are remarkable achievements of the human mind in demonstrating how its two types of symbolic information processing can be coordinated in a systematic model.

Notes

1. "數者，理之時也。辭者，數之義也。吉凶者，辭之斷也。惠廸從逆者，吉凶之决也。" Cai Chen, *Hongfan*, 2:21a.
2. A few are shared with the *Yijing* (蒙，訟，比，豫，晉，損，益，升，革）, *Taixuanjing* (中，守，成，戾，止，翕，親，閒，飾，養), and *Dongji* (華，平，用，止), and *Qianxu* (昧，育，賓). Note that 潜 and 虛, the two words in the title of 司馬光's system, are individually two of Cai's Number Names.

3. See Wu Cheng 吳澄 (1249–1333) *Yueling qishierhou jijie*. Dating of the *Liji* and the question of Han Dynasty redactions have long been debated (see 月令 article in zh.wikipedia for summary); however, the phrases also appeared in the first twelve sections of the *Lushi chunqiu* 呂氏春秋 (3rd cen., BCE.).
4. In addition to those in the Luck Schedule, these include Prodigious (*yuan* 元, though an alternate connotation for this character, "black," is used in 8.8 Weapons); Success (*heng* 亨); Profit (*li* 利); Purity (*zhen* 貞).
5. 今人須以卜筮之書看之，方得。Zhu Xi, *Zhuzi yulei*, *bushi* 卜筮 chapter, paragraph 7. See also Zhu Xi and Joseph Adler, *The Original Meaning of the Yijing*, 13.
6. c.f. "原始反終；故知死生之說。 What begins at the origin returns to the end; thus, one knows what is said about life and death." *Yijing*, *Xici shang*, paragraph 4.
7. c.f. *Yijing*, H1 *Qian*, *Wenyan*："九三曰：. . .知至至之，可與幾也。Nine in the third [line] says . . . one whose knowledge reaches the maximal can be conjoint with the incipient.". And *Yijing*, *Xici xia*, paragraph 5: "幾者動之微，吉之先見者也。君子見幾而作，不俟終日。 The incipient is the infinitesimal [spring] of movement, the prevision of Good Luck. The noble-minded person sees the incipient and acts; he doesn't wait for the end of the day."
8. *Qing* is used once in the *Yijing* as a term of good fortune—in H55 *Feng* 6.5: "There comes distinction; there will be congratulations and praise. Good Luck. (*laizhang you qingyu ji* 來章；有慶譽。吉). The term appears several times in the *Tuan* and *Xiang* commentaries. *Yijing*, H2 *Kun*, *Tuan*, "Noble-minded person . . . in the end receives congratulations (*zhunzi*. . .*naizhong youqing* 君子。。。乃終有慶)."
9. *Dongzhi*, the 22nd of the 24 solar terms falls in the second half of the 11th month according to the Absolute Beginning Calendar (*Taichuli* 太初曆), a combined solar and lunar (agricultural) calendar first put into effect in 104 BCE. In the Gregorian calendar, it is around December 22nd.
10. *Yijing*, H1 *Qian* 9.1, the second line of the classic text after the hexagram words, reads "Hidden dragon, do not act (*Qianlong wuyong* 潛龍勿用). Here, this phrase similarly appears in the second line of text, though without "dragon."
11. Although not printed with its matrix, elsewhere in Cai's book are the lines: "數曰：潛，藏也。勿用有攸往，陽微也。正靜吉，正而靜，所以吉也。君子藏器於身，待時而動；故無不利也。 The Numbers say: 'Hiding" means stored away. 'Do not go anywhere' means *yang* is infinitesimal. 'Propriety, quiet—good luck' means Good Luck derives from having propriety and being quiet. Thus, the noble-minded person secretes implements on his body and bides his time before acting. [Then] nothing does not profit。" Cai Chen, *Hongfan*, 2:10b.
12. *Xiaohan*, the 23rd of the 24 solar terms, which falls in the twelfth month of the agricultural calendar, around January 6 of the Gregorian calendar.
13. Cai's *Yueling* citations refer to geese four times—here and in 2.6, 7.4, and 8.2. Wu Cheng says of this instance: "鄉，向導之義。二陽之候，鴈將避熱而回，今則鄉北飛之，至立春後皆歸矣，禽鳥得氣之先故也。'*Xiang*' has the meaning of 'lead.' in the two *yang* seasons[, spring and summer], geese avoid the heat and return [north]. Now they lead toward the north and fly; when it comes to after Advent of Spring they all go back [south].

This is because birds receive foresight regarding the elements." Wu Cheng, *Yeuling*, *Xiaohan* 小寒 section.

14. c.f., "擊蒙；不利為寇，利禦寇 Strike Naiveté. No Profit in being a bandit; profit in resisting bandits." *Yijing*, H4, *Meng*, 9.5.
15. *Dahan*, the 24th of the 24 solar terms, which falls in the twelfth month of the agricultural calendar，around January 20th of the Gregorian calendar.
16. *Yijing*, H42 *Yi*, *Tuan* explains the phrase "profit in going somewhere (*li youyouwang* 利有攸往) in the hexagram words with *zhongzheng youqing*—the last four characters of this Legend. This is taken by Zhu Xi and in general by commentators to refer to the fifth line of H42, ䷩, which is a *yang* line "properly" placed in an odd-numbered line position and "centered" in the middle of the upper trigram. *Dong Kai*, ch. 7b, paragraph 11–12.
17. This is the almanac phrase from the preceding Number, Waiting 1.8.
18. *Yijing*, H38 *Kui* 6.3 and H57 *Xun* 9.5 have the same phrase "无初有終 *wuchu youzhong*."
19. A note in the *Caishijiuru shu*, 6:49b, reads: "'Chong' means 'peace;' the opposite of [8.5] Weapons (*Chong, he ye*; *yu rung xiangdui*. 冲，和也；與戎相對。)" The only other source that suggests this connotation is the 11th century CE rhyme dictionary *Guangyun* 《廣韻,》 which gives the equivalents "和也，深也。*Heye*, *shenye*, peace, deep" with reference to lines from a poem by Xiao Que 蕭慤 (6th cen. CE): "重明豈凝滯，無累在淵冲。" (Note that the character 冲 is commonly written "沖.") Given the first *Guangyun* definition, these might read: "How could the double lights [sun and moon] stagnate? They do not pile up in the abyssal peace." However, "abyssal deep" is equally plausible. Nevertheless, the *Zhongyong* links 中—visibly present as the right-hand element in 冲—and 和 in "喜怒哀樂之未發，謂之中；發而皆中節，謂之和。When happiness, anger, grief, and pleasure have not yet arisen, it is called the centered (*zhong*); when they have arisen and are tempered by the center, it is called at peace (*he*)". *Zhongyong zhangju*, ch. 1. Cai perhaps played on the visual pun and availed himself of the obscure meaning of 冲 to make this association.
20. c.f. "君子體仁，足以長人 The noble-minded person embodies benevolence and is fit to bring out the best in others." *Yijing*, H1, *Qian*, *Wenyan*.
21. c.f., "首出庶物；萬國咸寧 The leader stands out above the many beings; the myriad countries all are at peace." *Yijing*, H1, *Qian*, *Tuan*.
22. *Lichun*, the first of the 24 solar terms, which falls in the first month of the agricultural calendar, around February 4th of the Gregorian calendar.
23. "Prodigious Success; profit in purity" (*Yuanheng*; *lizhen* 元亨；利貞) are the hexagram words for *Yijing*, *Qian*, H1, and the phrase also appears in six other *Yijing* hexagrams. In *Kun* 坤, H2, it is interspersed with two other words—" Profit in the mare's (*pinma* 牝馬) purity")—in the same manner as in this Legend.
24. *Yushui*, the second of the 24 solar terms, which falls in the first month of the agricultural calendar, around February 19th of the Gregorian calendar.
25. This image occurs in *Yijing*, H32, *Heng*: 6.5：恆其德。貞。婦人，吉。夫子，凶。Enduring is one's virtue—sooth. Wife—Good Luck; husband—Bad Luck."
26. Cai Chen here uses the same almanac line as in 7.4 Weakness. Wu Cheng believed the line for this period should be "timely geese go north (*houyingbei* 候鴈北)." But, he wrote, "《月令》、《漢書》作鴻鴈北。鴈，知時之鳥。熱歸塞北，寒來江南，沙漠乃其

居也 The *Yueling* and *Hanshu* have this as 'swans and geese go north.' Geese are birds who understand time. When it is warm, they return north of the passes; when it is cold they come south of the River. The desert is actually their home." Wu Cheng, *Yeuling, Hanyu* 寒雨 section.

27. This phrase appears in the *Shiji*: "夫政不简不易，民不有近；平易近民，民必归之。When governing is not simple, and not tranquil, it is because there is nothing to draw the people near. Balance and tranquility draw the people near—the people must thereby return to one." Sima Qian, *Shiji*, *Lu Zhougong shijia* 鲁周公世家, ch. 33.
28. A note in the *Caishijiuru shu* reads: "*Yi* means tranquility (*Yi, ping ye* 易，平也)." *Caishijiuru shu* 6:49b.
29. The term *wenming* appears as an attribute of the early "sage" emperor Shun 舜 in the *Shujing*, where he is described as *junzhe wenming* 濬哲文明. Cai Chen glossed this phrase as "深沈，而有智；文理，而光明。 He is profound, and has wisdom; is cultivated, and shines brilliantly." Cai, *Shujizhuan*, ch.1, paragraph 6. Cai's first two phrases are similar to an *Yijing Tuan* commentary trope for connecting a hexagram's trigram structure to its name and/or wording. In the *Tuan*, "*wenming*" refers to the trigram *li* 離, associated, among other traits, with the sun and light. *Yijing*, H36 *Mingyi*, *Tuan* says: "Inside is design that is brilliantly displayed, and outside is softness that is compliant (*neiwenming er wairoushun* 内文明而外柔順). These terms describe the trigram structure of this hexagram, ䷣ , where the trigram *li* occupies the first three, or "inside" lines, and the trigram *kun*, which is "soft and compliant" by nature comprises the fourth through sixth (top) lines. If this description of inner and outer trigrams were applied to Cai's words in the Kinship Legend, the implied hexagram would be the invert of *Mingyi* with the trigram *kun* on its "inside" and *li* on its "outside." That is H35 *Jin* ䷢ , whose name is the same character as that of Number Name 4.3 Progress. c.f. 3.7 Uplifting and 4.6 Distinctness, where the translation is based on an alternate connotation of *wenming*.
30. *Jingzhe*, the third of the 24 solar terms, which falls in the second month of the agricultural calendar, around March 6th of the Gregorian calendar.
31. In Five Phase (*wuxing*) theory, wood is associated with springtime, which is the current season in Cai's almanac. For example, Wu Cheng's *Yueling* commentary for the solar term Rain Water reads, "Spring from the first belongs to [the Phase of] Wood (*chun shi shumu* 春始屬木)." Wu Cheng, *Yueling*, *yushui* 雨水 section.
32. *Mengzi*, *Jinxin shang* 盡心上, paragraph 21 reads: "其生色也，睟然見於面，盎於背，施於四體。The outward appearance to which [the virtues of the noble-minded person give rise] is clearness of eye visible in the face and in broadness of the back suffused through the four limbs."
33. *Yijing*, H42 *Yi* 益,, *Xiang* : "君子以見善則遷，有過則改 The Noble-minded person uses [the time of Increase] to see the good then evolve [toward it], or, having error, then modifies [his behavior]."
34. *Yijing*, H49 *Ge* 革,, 6.6: "君子豹變；小人革面 Noble-minded person transforms like a leopard; small-minded person alters his countenance."
35. This phrase appears in the *Shujing*, *lüxing* 呂刑 chapter, where in judging lawsuits one is advised to "察辭于差，非從惟從。Examine what is said for discrepancies, and what you

at first might not have gone along with, you may decide to follow." Cai's gloss takes this to show an unbiased, balanced perspective. See Cai Chen, *Shujing jizhuan*, ch. 6. The *Shujing* phrase is reversed here.

36. *Yijing*, H52 *Gen* 艮："艮其背，不獲其身；行其庭，不見其人。无咎。 Stop that one's back, do not catch that one's self; traverse that one's rooms, do not see that one's person—no error."
37. The *Yueling* has *xuanniao* 玄鳥 and Wu Cheng the variant *yuanniao* 元鳥 for "swallow," a synonym for Cai's *yiniao*. Wu Cheng, *Yueling*, 春分 section. Note that in Sickness 7.5 玄鳥 also appears.
38. *Chunfen*, the fourth of the 24 solar terms, which falls in the second month of the agricultural calendar, around March 21st of the Gregorian calendar.
39. *Yijing*, H61 *Zhongfu* 9:2 "鳴鶴在陰，其子和之。Crane calls in the shadows, its fledgling responds in harmony."
40. *Yijing*, H31 *Xian*, *Tuan* reads "二氣感應以相與。The two types of *Qi* [*yin* and *yang*] stimulate and respond to each other as they work together."
41. *Yijing*, H59 *Huan*, 9.5 has the line "[Like] streaming sweat is his great shout (*Huanhan chi dahao* 渙汗其大號). Zhu Xi takes the sweat as a metaphor for the way in which a ruler's commands go forth in all directions and cannot be recalled. Zhu Xi *Zuziyulei*, *Yijiu* 易九, *huan* 渙, paragraph 13.
42. *Yijing*, H2 *Kun*, says, "In the southwest, get a friend (*Xinan depeng* 西南得朋).
43. These two sentences appear in *Yijing*, *Xici xia*, paragraph 5.
44. *Qingming*, the fifth of the 24 solar terms, which falls in the third month of the agricultural calendar, around April 5th of the Gregorian calendar.
45. "*Wenming*" is here translated as it appears in *Yijing*, H1 *Qian*, *Wenyan*, which says: " [This hexagram's second line's] 'See dragon in the fields' means that designs of all under heaven are brilliantly displayed ('*jianlong zai tian*,' *tianxia wenming* 見龍在田，天下文明。)" c.f. the translation in 2.8 Kinship and the note there.
46. This phrase appears in the *Shujing Yiji* 益稷 chapter, where Cai's commentary equates *xian* 獻 with *xian* 賢, "a worthy" and continues: "黎獻者，黎民之賢者也。*Lixian* means worthies among the common people." Cai Chen, *Shujing jizhuan*, 1:37a.
47. *Caishijiuru shu* 6:54b has 平 for this character, which fits this context better than its homophone and occasional synonym, *ping* 評.
48. *Guyu*, the sixth of the 24 solar terms, which falls in the third month of the agricultural calendar, around April 20th of the Gregorian calendar.
49. The *Liji*, *Leji* 樂記 chapter, paragraph 17 reads: "天高地下，萬物散殊，而禮制行矣。Heaven is high, earth is below; the myriad things are distributed between in all their diversity. The designing of rituals evolved to reflect this."
50. The opening passage of the *Lunyu*, chapter 12:1 reads: "顏淵問仁。子曰：克己復禮為仁。 Yanhui asked about benevolence. The Master said, 'constraining the ego and returning to the rites makes for benevolence."
51. *Lixia*, the seventh of the 24 solar terms, which falls in the fourth month of the agricultural calendar, around May 6th of the Gregorian calendar.

52. This translation follows the *Caishijiurushu*, 50b: "*Zhang* means 'distinctness;' the opposite of [6.4] Dimness (*zhang mingye; yumeixiangdui* 章明也；與昧相對).
53. *Yijing*, H1 *Qian*, *Tuan* has the line "[The many] types of things flow through shapes (*Pinwu liuxing* 品物流形)."
54. *Yijing*, H2 *Kun*, *Tuan* has this line.
55. *Yijing*, H3 *Zhun*, *Tuan* "The movement of thunder and rain make fulness replete (*leiyuzhidong manying* 雷雨之動滿盈)."
56. *Xiaoman*, the eighth of the 24 solar terms, which falls in the fourth month of the agricultural calendar, around May 21st of the Gregorian calendar.
57. This phrase appears in *Record of Rites* (*Liji* 禮記), 祭統 chapter, section 26. It pertains to the traditional customs surrounding the summer ancestral sacrifice (*di* 禘).
58. "Supple grass" is grass that bends with the wind. The Western Han collection *Shuoyuan* 說苑 says: "夫上之化下，猶風靡草，東風則草靡而西，西風則草靡而東，在風所由而草為之靡。Those in high positions transform those beneath them as wind [affects] the supple grass: if the wind is from the east, the grass bends toward the west. If the wind is from the west, the grass bends toward the east. Wherever the wind comes from, the grass is made to bend." Liu Xiang 劉向, ed., *Shuoyuan, Jundao* 君道 chapter, paragraph 4. This word is also used in the next Number Name, where it is rendered "Revelry."
59. The *Caishijiuru shu* 6:50b gives *duo*多, "many," as an equivalent for *mi* 靡. The similar character *chi* 侈 occurs idiomatically with 靡 to mean "decadent." The same character *mi* appears in the almanac reading for 4.8 Bestowing, *micao si* 靡草死. Perhaps the two usages are related by the ease with which the supple grass is bent by the wind. Note that the *Shuoyuan* passage cited in the footnote there also refers to those in high (*shang* 上) and low (*xia* 下) positions, in that case the former transforming the latter. In Revelry, the two types are alike idle and at ease.
60. reading *xun* 殉 for its homophone *xun* 徇.
61. *Kaitian bidi* 開天闢地 is an idiom that means "unprecedented." Its source is in the account of the origin of material existence cited from the 3rd century CE writer Xu Zheng's 徐整 *Sanwuli ji* 三五曆紀 in Ouyang Xun, *Yiwen leiju* ch. 1, *Tianbu shang* 天部上 chapter, *tian* 天 section, paragraph 23. Xu said that originally heaven and earth were in their primal chaotic condition like a chicken egg (*Tiandi hundun ru jizi*天地混沌如雞子), and after 18,000 years, heaven and earth broke out (*Tiandi kaipi* 天地開闢). The *yang* material was clear and formed heaven; the *yin* material was turbid and formed earth (*yangqing weitian, yin shu wei di* 陽清為天，陰濁為地).
62. "Ripeness" translates *qiu* 秋, which means "autumn." Wu Cheng, *Yueling*, 小滿 section explains: "秋者，百穀成熟之期，此於時雖夏，於麥則秋，故云麥秋也。Autumn is the time when the hundred grains come to ripeness. In this case, though the time falls in summer, it is autumn [in this sense] for wheat. Thus, it says, "wheat attains ripeness."
63. "高辛氏有才子八人，世謂之「八元」Gao Xin [legendary ruler, also called Digu 帝嚳, 3rd millenium BCE.?) had eight people of great talent; the world calls them "the Eight Adepts." Sima Qian, *Shiji*, *Wudibenji* 五帝本紀 section, paragraph 22.
64. "舜賓於四門，乃流四凶族，遷于四裔，以御螭魅 When Shun [legendary emperor, 3rd millenium BCE.?] received guests at the four gates, he banished the families of the Four Evil Ones to the four to rule over hornless dragons and goblins." *Ibid.*, paragraph 23.

65. *Mangzhong* the ninth of the 24 solar terms, which falls in the fifth month of the agricultural calendar, around June 6th of the Gregorian calendar.
66. *Xiazhi* the tenth of the 24 solar terms, which falls in the fifth month of the agricultural calendar, around June 21st of the Gregorian calendar.
67. *Zhongyong* ch. 1: "道也者，不可須臾離也，可離非道也。是故君子戒慎乎其所不睹，恐懼乎其所不聞。One cannot depart from the Way for the briefest moment–if it can be departed from, it is not the Way. Therefore, the noble-minded person is guarded, anxious about what (s)he does not see, is fearful about what (s)he does not hear." The *Shuowen* 說文 dictionary, *mubu* 目部 notes the equivalence of 睹 and 覩.
68. This phrase also appears in 5.6 Subservience.
69. *Xiaoshu*, the eleventh of the 24 solar terms, which falls in the sixth month of the agricultural calendar, around July 7 of the Gregorian calendar.
70. *Yijing*, H62 *Xiaoguo* 小過,: "飛鳥遺之音不宜上；宜下 The sound left by a flying bird does not benefit that above; it benefits that below."
71. *Lunyu*, *Xianwen* 憲問 14:25: "夫子[蘧伯玉]欲寡其過而未能也。 My lord [Lian Boyi] desires to make few his errors and is unable."
72. These words come from the *Shujing*, *Hongfan* chapter that describe one's "five characteristics (*wushi* 五事)" and their ideal manifestations. This excerpt concerns the first of these: "The first is one's demeanor (*yiyuemao* 一曰貌); . . . one's demeanor should be respectful (*maoyuegong* 貌曰恭); . . . respect is shown in seriousness (*gongzuosu* 恭作肅)." These translations reflect Cai's gloss in *Shujizhuan*, ch. 4, paragraph 15.
73. The *Zhongyong* and *Daxue* both contain this phrase with *shen* 慎 instead of *jin* 謹. Both words mean "careful, prudent," the connotation in contemporary Chinese of the two used together, *jinshen*. Specifically, in the *Zhongyong*, "莫見乎隱，莫顯乎微，故君子慎其獨也 Nothing is more visible than what is hidden; nothing is more obvious than what is incipient. Thus, the noble-minded person is careful in his solitude." This is from the passage cited in 5.6 and 5.7. In the *Daxue*, "君子必慎其獨也 The noble-minded person must be careful in his solitude."
74. This sentence is from the *Zhongyong* passage: "其次致曲，曲能有誠，誠則形，形則著，著則明，明則動，動則變，變則化，唯天下至誠為能化。" Zhu Xi glosses, "Qu means an inclination (*qu yipian ye* 曲，一偏也), making the passage translate: " [Those lesser than the sages] cultivate incipient inclinations; from those inclinations can come sincerity. Once one has sincerity, it will be expressed by taking form; once formed, it will be manifest; once manifest, it will shine bright (*ming*); once shining bright, it will cause action; once causing action, it will make change; once making change, it will transform [for the better]. Only the utmost sincerity under heaven can thus transform." *Zhongyong Zhangju*, ch. 23. Note that the full passage also contains the "brilliant (*ming* 明) of Cai's following line."
75. The *Daxue*, ch. 2 has the passage: "*kemingde* 克明德" which, taking Zhu Xi's gloss "'Ke' means 'is able to' (*ke nengye* 克，能也) translates "is able to let shine one's virtue." *Daxue Zhangju*, ch. 2. Cai combines these allusions to evoke both the sense of "shining bright" in the previous note and the enlightenment that is the essence of sincerity.
76. *Han shi waizhuan*, ch. 2: "君子盛德而卑，虛己以受人，旁行不流，應物而不窮，雖在下位，民願戴之，雖欲無尊，得乎哉. The noble-minded person has abundant virtue

but is humble, empties oneself to receive others, adjusts to the time and is not carried away, and responds to things without ceasing. Though he be in a low position, the people want to support him; though he desires not to be honored, he is." "Empties oneself" is another association with this Legend.

77. *Dashu*, the twelfth of the 24 solar terms, which falls in the sixth month of the agricultural calendar, around July 23rd of the Gregorian calendar.
78. *Zhongyong* ch. 33: "君子之道，闇然而日章。In the Way of the noble-minded person, where things are murky, they daily become distinct."
79. *Yijing*, H36 *Mingyi*, *Xiang*: "君子以蒞眾，用晦而明。The noble-minded person in managing the masses obscures his enlightenment."
80. *Yijing*, *Kuei* 睽 H38, 6:9: "往遇雨，則吉 Go and meet rain, then Good Luck."
81. These sentences are adapted from *Lunyu*, *Liren* 里仁 4:16: "子曰:君子喻於義，小人喻於利."
82. *Liqiu*, the thirteenth of the 24 solar terms, which falls in the seventh month of the agricultural calendar, around August 8th of the Gregorian calendar.
83. A note in the *Caishijiuru shu* reads: "'Shi' means 'withdrawal' (*Shi*, *tui ye* 郤， 退也，" *Caishijiuru shu*. 6:51b. This suggests that Cai read 郤 as *que* 卻, which means "withdraw" and is still paired with 退 idiomatically in contemporary usage.
84. *Yijing*, *Xici*, *shang*, paragraph 6: "夫坤，其靜也翕，其動也闢，是以廣生焉。As for [H2] *Kun*, its calmness is agreeable, its movement is toward openness; thus, it broadly gives birth."
85. These lines are adapted from *Daxue* ch. 11: "財聚則民散，財散則民聚。"
86. *Chushu*, the fourteenth of the 24 solar terms, which falls in the seventh month of the agricultural calendar, around August 23rd of the Gregorian calendar.
87. Adapted from *Lunyu*, *Jishi* 季氏 16:8: "孔子曰:君子. . .畏天命; . . .小人不知天命而不畏也。Kongzi said, 'The noble-minded person . . . is in awe of Heaven's Command; . . . the small-minded person does not understand Heaven's Command and is not awed."
88. *Bailu*, the fifteenth of the 24 solar terms, which falls in the eighth month of the agricultural calendar, around September 8th of the Gregorian calendar.
89. Here the birds are flying north. Wu Cheng's *Yueling* commentary reads, "鴻《淮南子》作候鴈來。鴻大鴈小，自北而來南也。不謂南鄉，非其居耳。 Instead of 'swans' *Huainanzi* reads 'timely geese come.' Swans are large, geese are smaller. From the north they come southward. It doesn't say 'leads toward the south' "[as in 1.4 Extending] because [the south] isn't their home range." Wu Cheng, *Yeuling*, *Bailu* 白露section.
90. *Yijing*, H27, *Yi*, *Xiang* commentary ends with these three words.
91. *Qiufen*, the sixteenth of the 24 solar terms, which falls in the eighth month of the agricultural calendar, around September 23rd of the Gregorian calendar.
92. *Mengzi*, *Gaozi shang* 告子上 chapter, paragraph 11: "學問之道無他，求其放心而已矣。The Way of study is no other–it is only in seeking one's wandering mind."
93. *Yijing*, H23 *Bao*, 6.9: "碩果不食。Best fruit not eaten."
94. *Shujing*, *Hongfan* chapter: has this phrase in the context: "歲、月、日、時無易。百穀用成；乂用明；俊民用章；家用平康。If the years, months, days, and hours do not vary [from seasonableness], the hundred grains will mature; government policies will be

enlightened; the talented among the people will be distinguished; families will live in peace and health." Cai, *Shujijie* ch. 4, paragraph 15.

95. *Yijing*, H20 *Guan*, 6.4: "觀國之光；利用賓于王。 Observe the country's brightness; profit in having a sojourn with the king."
96. *Hanlu*, the seventeenth of the 24 solar terms, which falls in the ninth month of the agricultural calendar, around October 8th of the Gregorian calendar.
97. Wu Cheng's *Yueling* commentary reads: "賓，客也。先至者為主，後至者為賓，蓋將盡之謂 'Sojourn' means 'be a guest.' The earlier arrivals act as host; later arrivals are guests. This would seem to speak of nearing completion [of the migration]." Wu Cheng, *Yeuling, Hanlu* 寒露 section.
98. *Yijing*, H26 *Dachu*, *Tuan* uses these four words to describe that hexagram.
99. In Five Phase (*wuxing*) theory, metal is associated with autumn, which is the current season in Cai's almanac. For example, Wu Cheng's *Yueling* commentary for the solar term White Dew reads, "Autumn belongs to [the Phase of] Metal (*qiu shu jin* 秋屬金)." Wu Cheng, *Yueling, Bailu* 白露 section.
100. *Shuangjiang*, the eighteenth of the 24 solar terms, which falls in the ninth month of the agricultural calendar, around October 23rd of the Gregorian calendar.
101. This legend references several *Yijing* images of war. *Yijing*, H2 *Kun* 6.6: "龍戰于野。其血玄黃. Dragons war in the wilderness; their blood is black yellow." The *Wenyan* for this line reads "天玄而地黃。 *Tian xuan er di huang* Heaven is black and earth is yellow." 元 *yuan* can be used for 玄*xuan*, and that seems to be the intention in this context.
102. *Yijing*, H1 *Qian* 9.6: 亢龍 *kanglong* overreaching dragon."
103. *Yijing*, H7 *Shi*, 1.6: "師出以律，否臧凶。 Army goes out with regulations; if [the battle is] not for a good [cause]—Bad Luck." This translation follows Zhu Xi's gloss "'*Fouzang*' speaks of 'not for the good'" (*fouzang wei bushan* 否臧謂不善). Dong Kai, ch. 2b, paragraph 151.
104. *Lidong*, the ninteenth of the 24 solar terms, which falls in the tenth month of the agricultural calendar, around November 7th of the Gregorian calendar.
105. *Yijing*, H2 *Kun* 1.6: "履霜，堅冰至 Tread on frost, firm ice is coming."
106. This phrase contains allusions to two classical sources. *Yijing, Kan* 坎, H29 says:"維心亨 Bind up the heart—success." Cai replaces 維 with its homophone, 惟, which can mean "think about." This is the sense in which it appears in Mengzi: "孔子曰：'操則存，舍則亡；出入無時，莫知其鄉。' 惟心之謂與？ Kongzi said: 'Control it and it is preserved, set it aside and control is lost; it comes and goes at its own time, and there is no knowing its location,'—this speaks of contemplating the mind." *Mengzi, Gaozi shang* 告子上, paragraph 8.
107. *Yijing*, H27 *Yi*,: "觀頤，自求口實 Observe the mouth—it seeks its own fruits."
108. *Yijing*, H44 *Gou*, which the *Tuan* commentary says means "meeting," (*Gou, Yuye* 姤，遇也), reads, "女壯，勿用取女。The woman is strong; do not take a woman."
109. *Shiji* 《史記》 "*Qishijia* 齊世家" paragraph 2, has the story, "西伯將出獵，卜之曰：所獲非龍非彲，非虎非羆，所獲霸王之輔。於是果得太公於渭之南。Xibo [King Wen of Zhou], about to go hunting, had the turtle shell divined, and it said: 'It is not dragon, is not basilisk, is not tiger, is not bear that you will catch. What you will catch is an assistant

to a hegemon king.' In fact, he obtained Taigong [Jiang Ziya 姜子牙] on the south band of the Wei River." Jiang Ziya was a military strategist who ultimately helped King Wen's son, King Wu, found the Zhou Dynasty.

110. *Yijing*, H14 *Dayou* 6.9: "自天祐之，吉；无不利 Protection comes from heaven——Good Luck; nothing does not profit."
111. *Xiaoxue*, the twentieth of the 24 solar terms, which falls in the tenth month of the agricultural calendar, around November 22nd of the Gregorian calendar.
112. *Yijing*, H21 *Shihe*: "亨。利用獄 Success. Profit in bringing lawsuits."
113. *Ren* is the ninth of the ten *tiangan* 天干, used in counting cyclical phenomena like years and days. *Caishi jiuru shu* 6:53b: "*Ren* means 'crafty in speech'; it is the opposite of [1.5] Straightness. (*Renning ye; yi zhi xiangdui* 壬佞也；以直相對)." The term in this passage translated "conception (*ren* 妊)" contains this element.
114. *Daxue*, the twenty-first of the 24 solar terms, which falls in the eleventh month of the agricultural calendar, around December 7th of the Gregorian calendar.
115. "Good end (*lingzhong* 令終)" is from the poem *Jizui* 既醉in the *Shijing* 詩經 *Daya* 大雅 section. "昭明有融、高朗令終。令終有俶、公尸嘉告." James Legge's translation reads "May your bright intelligence become perfect,/ High and brilliant, leading to a good end!/ That good end has [now] its beginning:/ The personator of your ancestors announced it in his blessing." (ctext.org/book-of-poetry/ji-jui/zh#n16395). The last line echoes *Yijing*, H15 *Qian*: "亨君子有終 Success, Noble-minded person has end."
116. See Schulz and Thomas J. Cunningham, "Seasonal Structure."
117. Regarding the *Taixuanjing*, see Michael Nylan, *The Canon of Supreme Mystery*.
118. Cai Chen, *Hongfan*, *Yuanxu*, 1b.
119. See Richard. J. Smith, "Fathoming the Changes,"146 ff.
120. See, for example, the discussion of Lai Zhide's differences in interpretation with Zhu Xi and others in Schulz, *Lai Chih-te*, 127 ff.
121. q.v. The *Caishijiuru shu* in fact takes all pairs to be opposites in a graphic on pp. 6:49a-53a.
122. 有天地，然後萬物生焉。盈天地之間者，唯萬物；故受之以屯。屯者，盈也。屯者，物之始生也。物生必蒙，故受之以蒙。蒙者，蒙也。
123. 1) 原者，氣之始也。冲者，形之始也。中者，治之極也。用者，物之窒也。終者，事之畢也。2) 原者，仁之先也。用者，義之端也。公者，禮之閑也。戎者，智之刓也。中者，信之完也。3) 原者，近乎中也。伏者，遠乎中也。近者，進；而遠者，退也。近者，息；而遠者，消也。4) 原始，反終；故知死生之說也。Cai Chen, *Hongfan*, 2:11b.
124. Finalized by Dong Zhongshu 董仲舒 (179–104 BCE.) in his *Ju xianliang duice* 舉賢良對策, cited in Ban Gu *Bohutong delun* Ch. 8, *Qingxing* 情性 section, paragraph 2.
125. *Yijing*, *Xici shang*, paragraph 4.
126. H4 *Meng* is Cai's 1.6; H6 *Song* is 7.8; H8 *Bi* is 4.1; H16 *Yu*, 5.3; H34 (*Da*)*Zhuang*, 3.6; H35 *Jin*,4.3; H41 *Sun*, 6.5; H42 *Yi*, 4.5; H46 *Sheng*, 5.4; and H49 *Ge*, 8.5. 5.7 Excess, *Guo*, might be thought to correspond to either H28 (*Da*)*guo* or H62 (*Xiao*)*guo*.
127. 1.1, 1.2, 1.7, 2.1, 2.2, 2.5, 2.6, 2.8, 3.2, 3.3, 3.4, 3.5, 4.7, 5.9, 6.4, 6.5, 6.8, 7.5, 8.1, 8.2, 8.4, 8.8, 8.9, 9.1, 9.2, 9.4, and 9.9.
128. 升。元亨。用見大人。勿恤，南征吉。

129. 豫。利建侯，行師。
130. 不明乎數，不足以語象；不明乎象，何足以知數？二者可以相有，而不可以相無也。Cai Chen, *Hongfan, Yuanxu*, 2a.
131. Where *ming* appears 15 times, sometimes with alternate shades of meaning.
132. 大學之道，在明明德，在親民，在止於至善。Zhu Xi, *Daxue zhangju*, 1b.
133. 明，明之也。明德者，人之所得乎天，而虛靈不昧，以具眾理而應萬事者也。*i-bid.*, 1b.
134. 但為氣稟所拘，人欲所蔽，則有時而昏；然其本體之明，則有未嘗息者。故學者當因其所發而遂明之，以復其初也。*ibid.*, 1b.
135. By contrast, Cai only once uses *zhen* 貞, generally taken as a synonym for *zheng* 正 in the *Yijing* (e.g., in the *Tuan* commentary for H7 *Shi* 師). The character *zheng* itself is only used once in the Yijing's texts, in the hexagram words of H25 *Wu Wang* 无妄.
136. 元吉 appears in favor of 吉吉 in some of Cai's matrices, for example, 1.1, 1.2, and 1.7 but not in 1.3–1.6. Similarly, 凶凶 is sometimes replaced with *daxiong* 大凶 (e.g., in matrix 1.1 but not 1.2). Perhaps the inconsistency represents a later editorial lapse. The term 元吉 appears once in the legends, in 1.1 Origin.
137. "八數周流，推類而求五中；則平." Cai Chen, *Hongfan*, 2:21b.
138. "皇君，建立也。極猶北極之極，至極之義，標準之名。中立，而四方之所取正焉者." Cai, *Shujizhuan*, 4:21b.
139. "極建，則大本立；極明，則大用著。以之齊家，而家無不齊；以之治國，而國無不治；以之平天下，而天下無不平。" Cai Chen, Hongfan 1:12b.
140. 中。赫赫大明；耀彼四隣。君子持盈，小人毀成.。
141. "中者，不偏不倚，無過不及之名。庸，平常也。" Zhu Xi also cited Cheng Yi's gloss on the title as "不偏之謂中；不易之謂庸。中者，天下之正道；庸者，天下之定理。 Not biased is what is called 'the center'; not changing is what is called 'the constant,' The center is the righteous Way of all under heaven; the constant is the fixed pattern of all under heaven." Zhu Xi, *Zhongyong zhangju* chapter 1.
142. Cai's poem "Lamenting the Honorable Liu Wenan" ends 后世表章尊所学，四書功足許繼朱宣 When in later times they memorialize the throne to reverence what is studied; there will be sufficient merit in the Four Books to carry on Zhu's teaching." *Gushiciwang*, 挽刘文安公.

APPENDIX I

CAI CHEN'S *SONGSHI* BIOGRAPHY[1]

沈字仲默。少從朱熹游，熹晚欲著《書傳》，未及為，遂以屬沈。《洪範》之數學者久失其傳，元定獨心得之，然未及論著。曰：「成吾書者沈也。」沈受父師之托，沈潛反復者數十年，然後成書，發明先儒之所未及。其於《洪範》數，謂：「體天地之撰者《易》之象，紀天地之撰者《範》之數。數始於一奇，象成於二偶。奇者數之所以立，偶者數之所以行。故二四而八，八卦之象也；三三而九，九疇之數也。由是八八而又八八之為四千九十六，而象備矣；九九而又九九之為六千五百六十一，而數周矣。《易》更四聖而象已著，《範》錫神禹而數不傳。後之作者，昧象數之原，窒變通之妙，或即象而為數，或反數而擬象，牽合傅會自然之數益晦焉。」[2]

[Cai Yuanding's son] Chen's style-name was Zhongmo. When young, he followed Zhu Xi in his travels. In his later years, Xi wished to write a commentary on *Book [of History]* but could not do it; so he committed it to Chen. The number theory in the *Hongfan* [chapter] of the *Book [of History]* had long lost [the continuity of] its transmission. [Cai] Yuanding alone had gotten it in his mind, but he had not committed it to writing. He said, "The one who will complete my book is Chen." Deeply impressed by his father and teacher's trust, Chen retreated into reclusion for several decades and afterwards completed the book, discovering what earlier Confucians had not found. Regarding the numbers of the *Hongfan*," he said, "the images of the *Yijing* embody what heaven and earth engender;[3] the numbers of the [*Hong*]*fan* give temporal order to what heaven and earth engender. Number begins with one, odd; images are completed in two, even. Odd is that by which number is established; even is that by which number is put into action. Thus, two fours make eight—the images of the eight trigrams. Three threes make nine—the numbers of the [*Hongfan's*] Nine Divisions. From this, eight eights and again eight eights make 4,096, and the images are fulfilled. Nine nines and again nine nines make 6,561, and the numbers come full round. The *Yijing* was processed through the Four Sages, and its images were stated; the [*Hong*]*fan* was given to Sacred Yu, but its numbers were not transmitted. Later writers obscured the origins of images and number and impeded the mysteries of seamless transformation, either by taking images to be numbers or by reversing numbers and making them imitation images. [In either case], they dragged them together into forced association so that the numbers of nature were increasingly in the dark."

Notes

1. *Songshi*, 宋史 列傳193.
2. This quote is taken from Cai's preface to the *Hongfan huangji neipian*, though the last lines in the latter source contain variants as indicated with {}: "或{積}象以為數，或反數而擬象。{《洞極》有書，《潛虛》有圖，非無作也。} 而牽合傅會自然之數益晦{蝕}焉。" The first is insignificant, but the additional sentence alludes to two earlier divination systems, the *Dongji zhenjing*, attributed to Guan Lang and the *Qianxu* of Sima Guang. It reads, "The *Dongji* has the [*Luo*]*shu*, the *Qianxu* has the [River] Chart—these were worth producing." The last variorum means "eclipse" and intensifies Cai's sense of the darkness obscuring number.
3. This phrase appears in *Yijing*, *Xici xia*, paragraph 6.

APPENDIX II

BIBLIOGRAPHIC NOTE

Sikuquanshu zongmu zibuqi tiyao 四庫全書總目子部七提要
洪範皇極內篇
Master Plan Supreme Pivot Inner Writings
術數類一 數學之屬
Mathematics Section, Divination Category, One

臣等謹案，《洪範皇極內篇》五卷，宋蔡沈撰。沈父，元定，究心洪範之數；未及論著。嘗曰：「成吾書者，沈也。」沈反覆數十年，然後成書。分「外篇，」而釋數之辭，尚未備。故各條之下有但標「數曰，」二字，而無其文者。《永樂大典》及《性理大全》皆作《洪範內篇。》惟熊宗立注本以論三篇為「內篇」數八十一章為「外篇。」
考是書數八十一章擬易六十四卦當為「內篇；」論三篇擬《易》〈繫辭，〉〈說卦，〉等傳當為「外篇。」今各本皆以論三篇列於前，而八十一章列於後；倫序頗為不協疑《性理大全》與《永樂大典》同時纂輯所據同一誤本。未及詳考歟。
明俞深[1]著《洪範疇解，》曹溶稱為「釋蔡氏《內篇》疇即八十一章之數也。」程宗舜作《洪範內篇釋，》其自序曰：「釋八十一章之數,」亦不指三篇之論。韓邦奇引論中象以偶為用數語，作《洪範傳，》以別於經即外篇矣。意其時必有流傳善本與永樂間書局所據不同；故諸家之言如此；其訛似無可疑。然俞深等所據之本今不復見。未敢輕改古書姑仍其舊第編之。
又考王應麟《玉海》載此書，名《洪範數。》王圻《續通考》作《洪範皇極内外篇。》朱彝尊《經義考》作《洪範内外篇。》今詳考其書，當以《續通考》所名為是。《續通考》不載卷數；《經義考》作七卷。今以類相從編為五卷.
考洛書之名見於易，不見於書。《洪範》之文，以明理，非以明數。其事絶不相謀。後人以《乾鑿，》《太一行》九宮法指為洛書。案《史記》日者 〈列傳〉所載占日七家，太乙家居其一。《漢書》載 〈太乙諸術〉亦列於〈五行家。〉明為方技之說事，不出於經義矣。
盧辯注《大戴禮記》〈明堂〉篇，始附合於〈龜文。〉按盧辯，北齊人；其說最為晚出。朱子引此注，以證「龜書」指為鄭康成撰。朱傳及羣書，豈不知康成未註大戴禮記？特欲申龜文之說，别無古證。是不得不移之鄭康成耳。
至宋而圖書之說大興，遂以為洪範確屬洛書，洛書確屬龜文，龜文確為戴九履一等九數，而聖人叙彝倫之書變為術家談奇偶之書矣。
沈作是書，附會劉歆河圖，洛書，相為表裏八卦；九章相為經緯之說。借書之文以擬易之貌。以九九演為八十一疇，仿易卦八八變六十四之例也。取月令節氣分配八十一疇；陰用孟喜解易卦氣值日之術也。其揲蓍以三為綱，積數為六千五百六十一，陰用焦贛六十四卦各變六十四卦之法也。大意以《太元、》《元包、》《潛虛》既已擬易；不足以見新奇技變幻。其說歸之〈洪範〉實則朝三，暮四；朝四，暮三同一僭經而已矣。此在術數之家，已為重儓之。重儓本不足道。
以自沈以後，又開洪範之一派。支離轇轕，踵而為之者頗多。既有其末，不可不著其本。故録而存之，而别録於術數類。明非說經之正軌，儒者之本務也。
乾隆四十六年九月恭校上

Your several servants cautiously observe: *Master Plan Supreme Pivot Inner Writings (Hongfan Huangji neipian)* in five chapters was written by Cai Chen. Chen's father, Yuanding, focused his mind on the numbers of the *Hongfan* but didn't set forth his ideas in writing. He said, "The one who will complete my book is Chen." Chen returned to it frequently for several decades, and at last completed a book. He proposed to divide out an "Outer Writings" portion to explain the words given to the numbers, but it is not complete. Therefore, below each line of wording there is only the title "[The] Numbers say"—those two words and no associated text. *Yongle Dadian* [an encyclopedia published in 1408,] and *Xingli Daquan*[a digest of Confucian philosophy published in

1415,] both call the book *Hongfan neipian*. Only the edition annotated by Xiong Zongli [1409—1482] takes the three essays of theoretical discourses to be the "Inner Writings" and the 81 sections of Numbers to be the "Outer Writings."

Our view is that this book's 81 sections of Numbers that imitate the *Yi*'s 64 hexagrams should be taken as the "Inner Writings" and the three discursive essays imitative of the *Yi[jing]*'s *Xici*, *Shuogua*, and other commentaries to be the "Outer Chapters." The various current editions use a standard order placing the three discursive essays at the beginning and the 81 Number sections afterwards, and this does not especially help dispel the doubt that the *Xingli Daquan* and *Yongle Dadian* may have relied on the same corrupt edition to edit their own manuscripts. But this cannot be more minutely examined.

In the Ming Dynasty, Yu Chen [*jinshi* 進士, 1475] wrote *The Master Plan Divisions Explained* (*Hongfan choujie*), which Cao Rong [1613—1685] stated was an explication of Mr Cai's "Inner Writings" in which the "Divisions" were the 81 sections of Numbers.[2] Cheng Zongshun [fl. ca. 1550] wrote the *Master Plan Inner Writings Explication* (*Hongfan neipianshi*), and his personal preface says it "explicates the 81 sections of Numbers"; and it does not mention the three discursive essays. Han Bangqi [1479—1555] cited several phrases from the discourses, like "images take opposite pairings for their application,"[3] and he wrote about them as coming from the *Hongfan*'s "commentary" as opposed to from [its] "classic" [portion]—this ["commentary"] refers to the "Outer Writings." We think that at the time there must have been a rare copy in circulation different from the one that the publishers of the *Yongle* period [1403–1424] relied on; thus, the words of the various writers are such as we have noted, and their errors cannot be doubted. The text that Yu Chen and the others relied on is now no longer seen, but we dare not lightly change old books, and we continue to present that old material in our edition.

Wang Yinglin [1223–1296] in his *Yuhai* lists this book's name as *Master Plan Numbers*. Wang Qi [1530–1615] in his *Xuwenxian tongkao* says, "Master Plan Supreme Pivot Inner and Outer Writings." Zhu Yizun [1629–1709] in his *Jingyi kao* has *Master Plan Inner and Outer Writings*. Now minutely examining the matter, this book should take the title given in *Xutong kao* as correct. *Xutong kao* does not list the number of chapters; *Jingyi kao* says 7 chapters. In keeping with the edition we follow, we make it 5 chapters.

We observe that the name "*Luoshu*" is seen in the *Yi[jing]* and it is not seen in the *Shu[jing]*. The language of the [*Shujing*'s chapter] *Hongfan* elucidates principles; it is not for elucidating numbers. The [two] matters are decidedly not compatible. Later people pointed to the "nine palace" method used in the [apocryphal] *Qianzao* and *Taiyixing* [compass] as being the *Luoshu*. We observe that the *Shiji*'s "Biographies of Astrologers" contains seven types of astrologers and the *Taiyi* practitioners constitutes one among them. The *Hanshu* places the various *Taiyi* arts among the Five Phases (*Wuxing*) practitioners. Clearly these sayings [of Cai's] belong to wizardry and do not come from the meaning of the Classics.

In his commentary on the *Dadai liji*, "*Mingtang*" chapter, Lu Bian [ca. 520 CE] first linked [the *Luoshu*] to the [divinatory] turtle markings. We observe that Lu Bian was a man of northern Qi; so his words emerged very late. Zhu Xi alluded to this gloss in order to give evidence for the "turtle markings," but he pointed to it as having been written by Zheng [Xuan 玄, styled] Kangcheng [127–200 CE]. Zhu Xi annotated a great number of books—how could he not have known that Kangcheng didn't write a commentary on the *Dadai liji*? He particularly wanted to advance his theories on the turtle markings, and there was no other ancient evidence. So he could not but ascribe it to Zheng Kangcheng.

In the Song, [*He*]*tu* and "[*Luo*]*shu*" theories abounded, and it followed that the *Hongfan* must comprise the *Luoshu*, the *Luoshu* must comprise the turtle markings, the turtle markings must comprise the 9 numbers of "carry nine and tread on one,"[4] etc. And the eternal principles transcribed by the sages were transformed into a numerologist-speak "odd-even" [fortune telling] book.

In writing this book, Chen amalgamated the *Hetu* and *Luoshu* of Liu Xin [sic][5] in which to fit the eight trigrams and the nine sections [of the *Hongfan*] to make apocryphal theories. He borrowed the *Shu*[*jing*]'s words to imitate the *Yi*[*jing*]'s style and by extending nine times nine makes 81 divisions, he aped the idea of the eight *Yi*[*jing* trigrams] times eight transforms to 64. He took the *Yueling*'s seasonal segments and parceled them among his 81 divisions, shadowing the technique Meng Xi applied when explicating the *Yi*[*jing*] with the *Guaqi* assignment of hexagrams to the days [of the year]. His method of sorting yarrow stalks using three as the heading and building it up to 6,561 [matrix cells] shadows Jiao Gan's method of 64 gua each changing to 64 gua [in the *Yilin*]. In general, in that the *Taixuan*, *Yuanbao*, and *Qianxu* had already imitated the *Yi*[*jing*], there is nothing here that one would see as newly unique. Altering what those [earlier *Yijing* imitators] said and referring it back to the *Hongfan* is just three in the morning four in the afternoon, four in the morning three in the afternoon[6]—they all were alike in merely usurping the classics.

There is no need saying that this is a "slave of a slave" volume among the numerologists' [texts].

After Chen, there began a *Hongfan* School that developed in complex variant offshoots and became very prolific. Because we have this tree, we cannot but note its root. Thus, we record and preserve them and otherwise list them in the divination category. Clearly [Cai's approach] is not the proper channel for discussing the classics, which is a Confucian's basic duty.

Respectfully offered up in the ninth month of the forty-sixth year of the *Qianlong* reign [1781].

Additional Publication Information

Yao Yong's postface in the *Sikuquanshu* edition, 5.24a, says that he worked from a *Zhanggong jun zhai* edition 章貢郡齋刊本 in appending two graphics. The publisher *Zhanggong jun zhai* issued several books by Zhu Xi in the early 13th century.

Cai Chen's book appears with numerous variations in Cai Xuesu, ed., *Writings of the Nine Confucians of the Cai Family (Caishijiuru shu)*, 18a–57b. Some of the graphics that follow Cai's preface in the *Sikuquanshu* edition appear in the middle of the first discourse (篇上) and others appear after the third discourse (篇下). Several of the latter do not appear in the *Sikuquanshu* edition, and the two graphics with square matrices surrounded by circular arrays that are appended to the *Sikuquanshu* edition are also unique to the *Caishijiuru shu*. These are followed by a list of Number Names that gives a synonym for each and notes their opposite by name on pages 49a–53a. Then the Luck Spectrum is listed, also with one-word synonyms, and, last, the legends. There are no annotations and the 81 matrices are nowhere shown.

Notes

1. or 琛.
2. The *Qinding gujin tushujicheng* 欽定古今圖書集成 of 1726 reported his book to be "lost (*shi* 佚)." Volume 581， 怕。16b.
3. 洪範傳曰：「象以偶為用者也」 in Han Bangqi, *Hongfan tujie*, ch. 1, paragraph 15.
4. citing the mnemonic for the location of numbers in the *Luoshu* (of unknown date, still used in *Fengshui* 風水 practice): "戴九，履一；左三，右七；二四為肩；六八為足；五居中央。 Carry nine, tread on one; left three, right seven; two and four make the shoulders; six and eight make the feet; five resides in the center." See 河圖洛書 article, zh.wikipedia.org.

5. *Xin* 歆 should be *Mu* 牧 with reference to Liu Mu 劉牧. See Chapter 1, note 4.
6. citing the anecdote in *Zhuangzi* 莊子 *Qiwulun* 齊物論: "狙公賦芧，曰：「朝三，而莫四。」眾狙皆怒。曰：「然，則朝四，而莫三。」眾狙皆悅。 A monkey-keeper giving them chestnuts to eat said, 'three in the morning and four in the afternoon.' The monkeys were all angry. So he said 'All right, then—four in the morning and three in the afternoon.' And the monkeys were happy." The *Sikuquanshu* editors use this anecdote to mean "the same thing in a different form."

APPENDIX III

THE "HONGFAN SCHOOL"

As with the *Yilin*, the *Taixuanjing*, the *Dongji*, and the *Qianxu*, there is scant record of Cai's *Master Plan* being put to use as a divination system. These inventions tended to devolve into intellectual curiosities, critiques of them occupying a late chapter of a conspectus like Huang Zongxi's 17th-century *Yixue xiangshulun* or their visualizations inserted at the ends of graphic anthologies.[1]

The *Sikuquanshu* editors' critique of Cai's *Hongfan huangji neipian* notes that it gave rise to a "Hongfan School (*Hongfanpai* 洪範派)." A few who might be included under this rubric sought faithfully to explicate Cai's discourses—frequent mention has been made above of Xiong Zongli's Ming Dynasty commentaries, and Zhang Pan 章懋 (Depan 德懋, 1437–1522), who will also figure in the discussion in Appendix IV as having deciphered Cai Chen's divination system. But most scholars in that "school" were united more by the possibilities for independent thinking opened by Cai's innovative use of the *Luoshu* and its nine numbers than by adherence to his approach. They viewed the *Master Plan* as an "unfinished book (*weijingzhishu* 未竟之書)," in the words of Liu Shiqu 劉世衢 (18th cen.), and sought to "finish" it after their own inclinations. Li Chunjin 李純謹 (fl. ca. 1518) added wording for each Number Name after Cai's words "the Numbers say" and wrote commentaries in the style of the *Yijing's Xugua* and *Zagua*. Han Bangqi devised additional graphics and alternate divination methods. Similarly, Cheng Zongshun 程宗舜 (fl. ca. 1550) wrote new sublegends to be used like the *Yijing's* changing lines with yet another variety of divination practice. None of these enthusiasts seems to have put the 81 matrices to the use Cai intended.

There is much of interest about the Hongfan School. Their alternate schemes for graphic display, for example, evolve from Cai's into a panoply of base nine number theory that has yet to be fully explored. Similarly, their ongoing desire to invent new stalk sorting systems is a testament to the creative latitude that predecessors like Cai had inspired. Not only those of the Hongfan School, but Image and Number scholars in general tended to believe that, as Lai Zhide said of his 16th-century *Yijing* commentary, "From the passing of Kongzi, the *Yi* has been lost until this day."[2] The freedom of such iconoclasts to exercise originality with respect to that most revered of the Confucian Classics and to scorn the work of their predecessors speaks to the outlet for independent thinking that the *Yijing* provided in general.[3] As long as one could justify his ideas with references from classic literature, he could diverge from orthodoxy with the only risk being the kind of criticism the *Sikuquanshu* editors level at Cai—that he had misconstrued the classical sources or forced his interpretation to advance a personal theory. Even Zhu Xi came in for rallying on counts like the questionable reference to a Han Dynasty scholar in the *Sikuquanshu* entry on Cai's book (see Appendix II). Nonetheless, Zhu Xi was the preeminent Neoconfucian, and when he sanctioned the graphics of Shao Yong as raw material for *Yijing* interpretation, he also assented to the establishment of parallel lines of study like the Image and Number and Hongfan Schools.

Notes

1. The miscellany at the end of editions of Lai Zhide's *Zhouyi* commentary with the title *Zhouyi Liazhu tujie* 《周易來注圖解》 (first appearing in a 1688 edition) is a good example. See Schulz, *Lai Chih-te*, 252–253.
2. Lai Zhide 來知德. *Zhouyi jizhu* 周易集注 (四庫全書 edn., archive.org), 原序, 2b.
3. Richard Smith comments on this in "Fathoming the Changes," 147.

APPENDIX IV

HONGFAN DIVINATION METHOD

Cai Chen prefaced his sketch of a divination method saying, "筮者神之所為乎 Sortilege—is it not what the spirits do?"[1] There are three explanatory sources for Cai's approach. The *Sikuquanshu* edition gives two glosses, one presumably by the general annotator Xiong Zongli and a second that attributes the "understanding" of the final numerical reading to a Ming Dynasty scholar Zhang Pan 章懋 (t. Depan 德懋, 1437—1522).[2] The third is a 17th-century description in Huang Zongxi's *Yixue xiangshu lun* that synthesizes the earlier sources.[3] The latter depended on Zhang Pan's solution to the mystery posed by point's seven through nine in Cai's description below. In essence, the methodology modifies the yarrow stalk sorting method associated with the *Yijing* to produce two bipartite numbers—one to establish a "big number (dashu 大數)" or "heading (*gang* 綱) equivalent to a Number Name, and one to find a "small number (*xiaoshu* 小數) or "aspect (*mu* 目) for the applicable pair of luck terms from the matrix of that Number Name.

Cai's original description reads:

1. 其著五十。虛一；分二；掛一 The number of stalks is 50. Remove one; divide the rest in two; hang one.
2. 以三揲之Sort the stalks in the two piles by threes.

3. 視左右手歸餘於扐 Observe the left and right hand [piles] and return the remainders as the held stalks.
4. 兩奇為一 Two odds make a one.
   Comment: as Huang Zongxi's reconstruction below elaborates, in this and the next two points, the process shifts from the counting off of point 3 that leaves a numerical remnant to treating the remnant as either odd or even in number.
5. 兩耦為二 Two evens make a two.
6. 奇耦為三 An odd and an even make a three.
7. 初揲綱也再揲目也 The first sort is the heading; the second sort is the aspect.
8. 綱一函三以虛待目 A heading count has three possibilities; [its outcome must] await in emptiness the aspect.
9. 目一為一以實從綱 An aspect of 1 makes 1 in order to realize what follows the heading.
   Comment: This is the problematic step that Zhang Pan deciphered and Huang Zongxi expanded in his treatment below. In essence, the "heading" is decided from several possible outcomes resulting from the first sort by the outcome of the second sort, which produces the "aspect."
10. 兩揲而九數具 Two sorts and a count of [one to] nine is complete.
11. 八揲而六千五百六十一之數備矣[4] Eight sorts and the 6,561 numbers are fulfilled.

The terms "divide," "hang," "return," "remainder," and "held" and the general outline of Cai's divination method come from the *Yijing*.

> The number of the great unity is 50, it's use is 49. Divide those into two [piles] to image two, hang one to image three, sort them by fours to image the four seasons, return to the odd in order to hold it to image the intercalary month. Every five years there is an additional intercalary month; so repeat [the process], hold [the remainder] and afterwards hang [them].[5]

From this laconic text, later scholars attempted to parse a methodology for *Yijing* divination by dividing of yarrow stalks, a form of cleromancy. Its terminology is controversial,[6] and the English translation here is a compromise drawing from the earliest sources collected in Kong Yingda 孔穎達, *Zhouyi zhengyi* 周易正義 and Li Dingzuo 李鼎祚, *Zhouyi jijie* 周易集解. Zhu Xi built on their understanding and finalized the process in his *Zhoui qimeng* 周易

啟蒙 chapter on *shifa* 筮法.[7] That explication was carried in the orthodox Confucian texts that were required reading for the civil service examinations.[8]

As a disciple of Zhu Xi, Cai Chen no doubt had the same understanding of the terms as his teacher. Beginning by removing one of the original 50 stalks, he sorts the remaining 49 into two piles and "hangs" one. Presumably, this "hanging" one is parked between the little and ring fingers of the left-hand as in Zhu Xi's instructions, but Cai does not specify that.[9] Zhu Xi then sorts the two piles, left then right, by fours. But Cai's numerical system is based on multiples of three, and he sorts by three rather than four. In both systems, the remainder must be nonzero; so Zhu Xi would see remainders of one, two, three, or four, and Cai would end up with one, two, or three. The remainders of both piles are "returned" to the hand holding the first "hanging" stalk and "held" between the ring and middle fingers and middle and index fingers. In Zhu Xi's method the stalks in the left-hand—not counting the "hanging" stalk—are summed up and recorded, and after three such operations a number is derived that is still subject to a conversion process before yielding one of a hexagram's six lines.

In Cai's approach, the left and right remnant bundles appear to be still treated separately with the "hanging" stalk counted along with the remnant bundle from the left-hand pile. This can be inferred from the descriptions of his method by Xiong and Huang, which allow for a result of 4 in the "held" count. A four, it seems, could only be derived by adding the "hanging" stalk to one result of the sorting process. A left-hand remnant of three, held next to the "hanging" stalk, seems the likely source for a count of four.[10]

Huang Zongxi detailed his understanding of Cai's process,[11] as translated below with the same outline numbering as in Cai Chen's description above. In points 4 and 5, he follows Xiong and in points 8 11 expands on Zhang Pan's "discovery" of the dependency of the left-hand number of a heading aspect pair on the result of the right-hand number.[12]

1. 取右刻一策，掛於左手小指間 From the right cutting take one stalk and "hang" it next the left-hand's little finger.
2. 以三揲之Sort the stalks [in the two piles] by threes.
3. 視左右手，歸餘於扐；取左刻之策以三數之餘：或一、或二、或三。歸扐於左手無名指間。次揲右刻之策餘幾，歸扐於左手中指間。連掛左二，則右必二；左三，則右必一；左四，則右必三。是為一揲為綱 Observe the left and right-hand [piles] and return their remainders as the held stalks. Take the remainder from the stalks in

the left cut sorted by threes—one or two or three—and return them to "the held" between the left-hand's no name [ring] finger [and little finger]. Next return the right-hand cut's several remainder stalks to "the held" between the left-hand's middle finger [and ring finger]. Including the "hanging" stalk, if the left is two stalks, then the right must be two; if the left is three, the right must be one; if the left is four, the right must be three. This makes one sort and constitutes the heading.

4. [13]視左右之餘策：兩奇為一，左三右一是也 Observe the left and right remnant stalks: two odds make one—this refers to left three and right one.
5. 兩偶為二：左二右二是也 Two evens make two—this refers to two on the left and two on the right.
6. 奇偶為三：左四右三是也 Odd, even make three—this refers to four on the left and three on the right.
7. 範分左右，猶卦分上下。此雖有一二三之數，然未成一旁 This way of dividing into left and right is like the [*Yijing*'s] hexagrams division into upper and lower [trigrams]. Although there are the counts of one, two, or three, this does not yet constitute a pair.
   需之再揲；綱必待於目也。復合見存之策，除去掛扐之策。初揲得四三，則見存四十二；初揲得三一或二二，則見存四十五。分掛揲歸如前法，是為再揲，為目。掛扐之數，左四，則右必二；左三，則右必三；左二，則右必一。亦以兩奇為一左三右三是也。兩偶為二：左四右二是也。奇偶為三：左二右一是也 One must sort again; so the heading must wait for the aspect [for its number to be decided]. Regroup the stalks remaining after the "hanging" and "held" are removed. If the first sort resulted in four and three, one sees that what is left is 42 stalks. If the first sort resulted in three and one or two and two, one sees that what is left is 45 stalks. Divide these, and "hang," "sort," and "return" in the previous manner. This makes the second sort and makes the aspect. If the "hanging" and "held" count is four on the left, then the right must be two; if left is three, then right must be three; if left two, then right must be one. Again, one takes two odds to be a one; this is the meaning of "left three right three." Two evens make a two; this is the meaning of "left four, right two." Odd, even makes a three; this is the meaning of "left two, right one."
8. 綱一數具三數；故一可以為一，亦可以為二，亦可以為三。待目而分 The single number of the heading carries the possibility of three

numbers. That is, [a heading of] one can make [a heading of] one; it can also make [a heading of] two; it can also make [a heading of] three. One makes a determination among these possibilities depending on the aspect.

9. 二之為四五六，三之為七八九亦然。如初揲一，再揲一，則實其函之一；於左方立𝍠。再揲二，則實其函之二；於左方立𝍡。再揲三，則實其函之三；於左方立𝍢。如初揲二，再揲一，則實其函之四；於左方立𝍣。再揲二，則實其函之五；於左方立𝍤。再揲三，則實其函之六；於左方立𝍥。如初揲三，再揲一，則實其函之七；於左方立𝍦。再揲二，則實其函之八；於左方立𝍧。再揲三，則實其函之九；於左方立𝍨。此正九數也。That two can make four, five, or six and three can make seven, eight, or nine is also the case. If the first sort is one and the second sort is one, then [the heading] is realized as a one, and on the left place a "𝍠." If the second sort is two, then [the heading] is realized as a two, and on the left place "𝍡." If the second sort is three, then [the heading] is realized as a three, and on the left place "𝍢." If the first sort is two and the second sort is one, then [the heading] is realized as a four, and on the left place "𝍣." If the second sort is two, then [the heading] is realized as a five, and on the left place 𝍤."If the second sort is three, then [the heading] is realized as a six, and on the left place "𝍥." If the first sort is one and the second sort is three, then [the heading] is realized as a seven, and on the left place "𝍦." If the second sort is two, then [the heading] is realized as an eight, and on the left place "𝍧." If the second sort is three, then then [the heading] is realized as a nine, and on the left place "𝍨." In this way one gets the proper nonary number [of the left member of a header.aspect pair].

Table A.1: Aspect Conversion of Heading

綱 Heading→	1	2	3
目 Aspect↓			
1	1	4	7
2	2	5	8
3	3	6	9

10. 兩揲，而九數具；四揲，而數名立。兩揲具範之左方，復合四十九策，如前兩揲以具右方。此之九數也。如正九是一數，之九是二數，為一之二潛。如正九是五數，之九是六數，為五之六伏 Two sorts and a nonary number [or "proper number" of the left] is fulfilled; four sorts and the Number Name is established. Two sorts have been divided into left-hand numbers; then the 49 stalks have been regrouped; and, as before, two sorts fulfill the right-hand number [of the heading.aspect pair]. These latter make the "belong" nonary numbers. Thus, if the "proper" nonary number is a one, and the belong nonary is a two, that makes [the Number Name] two that belongs to one [1.2], Hiding. If the proper nonary number is a five, and the belong nonary number is a six, that makes [the Number Name] six that belongs to five [5.6], Concealment.
11. 八揲而六千五百六十一之數備。 Eight sorts and the 6,561 numbers are complete.

The next procedure, which is not specified in Cai's text but is explained in a gloss,[14] is to derive two more numbers from one to nine that map to a cell in the matrix for the Number Name established in the first round of stalk sorting. In this manner, two terms of the Luck Schedule are identified to give a projected trend of fortune. Huang Zongxi calls this the "changing nonary (*bianjiu* 變九)"—a term that appears in neither Cai's text nor its commentary. Huang describes its derivation in the following.

> 數名既定，復兩揲以具左方。是變九之正九也 The Number Name being fixed, perform two more sorts to fulfill the left place, which is the proper nine of the changing nine.
>
> 復兩揲以具右方；是變九之[之]九也 Perform two more sorts to fulfill the right place, which is the belong nine of the changing nine.
>
> 凡四揲而得大數下之小數 In all, with four sorts one gets the big number's subordinate small number.
>
> 如變正九是三，變之九是八，即為三之八 If the changing proper nonary is three and the changing belong nine is eight, the resu[lt is] [matrix cell] eight that belongs to three [3.8].

> 如大數得一之一為原，當於原下尋小數。大數下之橫即小數之左；縱即小數之右。橫三縱八相合之處便是變九也 If one gets 1.1 for the big number, that is [the Number Name] Origin. One must then [in the matrix] under [the Number Name] Origin seek the small number. The columns in the big number [matrix] are the small number's left terms; the rows are the right terms. The cell where column three and row eight intersect is the changing nonary [, in this case Blessing Felicity (*xiangxiu*), CH in the alpha equivalent employed herein].
>
> 一大數之下有八十一小數；以八十一乘之，得六千五百六十一數。吉凶休咎覩矣 Under one big number there are 81 small numbers. If one multiplies by 81, the product is 6,561, by which the numbers' Good luck, Bad luck, Felicity, and Error are seen.

In summary, the complexity of this system, though unnecessary for generating four random numbers, would have been intelligible, at least as Huang Zongxi described it, to a user familiar with Zhu Xi's stalk sorting concept, and it improved on Zhu Xi's by producing the input required to access Cai's divination readings with equal probabilities for all numbers. It translated the *Yijing* emphasis on the number two—the binary left and right-hand sorts that identified the 64 hexagrams—into three, square root of Cai's nonary system, and mapped directly to his matrices. It also provides a glimpse into Cai's thinking about numbers, which here influence one another in retrospect as the "aspects" define the "headings." This rather extraordinary concept suggests that the future mediates the present through some sort of numerical resonance, which would be analogous to the manner in which in Cai's matrices any given cell opposes and complements other cells in counterposed cells of its own matrix and across the suite of 81.

Notes

1. Cai Chen, *Hongfan*, 2:20a.
2. As Zhang Pan's dates are later than Xiong's and those of Yao Yong, who wrote a postface to Cai's book (see Appendix II), this reference must have been added in *Sikuquanshu* edition.
3. 4:43a–46b.
4. Cai Chen, *Hongfan*, 2:20 a-b.
5. 大衍之數五十；其用四十有九。分而為二，以象兩；掛一，以象三；揲之以四，以象四時；歸奇於扐，以象閏。五歲再閏，故再扐，而後掛。*Xici*, *shang*, paragraph 9.

6. Huang Zongxi discusses variant theories, 2:55a to 60b, and controversy has continued, as, for example, in Shih-chuan Chen "How to Form a Hexagram and Consult the I Ching, (*Journal of the American Oriental Society* 92:2, 1972), 237–249.
7. In this Zhu Xi was circumspect. As to whether the early Zhou Dynasty royals Wen Wang 文王 and Zhou Gong周公 used the method he proposed, he said, "他又須別有法，只是今不可考耳。且如《周禮》所載，則當時煞有文字。如今所見占法，亦只是大概如此。其間亦自有無所據底；只是約度如此。大抵古人法度，今皆無復存者 They most likely had another method; it is only that it cannot be researched now. Moreover, as with [the methods] the *Zhouli* mentions, at that time they had no words—as with some of the divination methods seen today. It is only that [the method I suggest] was in general like this, and within it there are unconfirmed aspects—it is only approximately like this. For the most part, there is now no way of recovering the methods of men of old." *Zhuzi yulei,Yi 2* 易二, *Bushi* 卜筮, paragraph 35.
8. See Zhu Xi and Adler, *Original Meaning*, 317–322. Also Wilhelm, *I Ching*, 721–724.
9. Nor does he say from which pile, left or right, the stalk is taken. Huang Zongxi notes that "孔氏[穎達]取左；朱子取右。 Kong Yingda (574–648) took from the left pile; Master Zhu [Xi] took from the right." *Yixue xiangshulun*, 2:55a.
10. Xiong's gloss in parenthesis follows Cai's original: "[item 4 above] 兩奇為一Two odds make a one (Xiong: 初揲三一，再揲三三 the first sort gives 3 and 1, the second sort [of the right-hand pile] gives three and three). [item 5] 兩耦為二 Two evens make a two (Xiong: 初揲二二，再揲四二 the first sort gives two and two, the second sort gives four and two).[item 6] 奇耦為三 an odd and an even make a three (Xiong: 初揲四三，再揲二一 the first sort gives four and three, the second sort gives two and one)." Cai Chen, *Hongfan*, 2:20a.
11. Huang Zongxi, *Yixue xiangshulun*., 4: 43a–46a
12. Zhang Pan's "discovery" is described in a gloss to Cai Chen, *Hongfan*, 2:20 a–b and translated in the following two notes.
13. Zhang Pan's gloss for points 4–6 in the *Sikuquanshu* edition of Cai's book read: 4.) 初揲三一；再揲三三。the first sort gives three and one; the second sort [of the right-hand pile] gives three and three. 5.) 初揲二二；再揲四二。The first sort gives two and two; the second sort gives four and two. 6.) 初揲四三；再揲二一。 The first sort gives four and three; the second sort gives two and one. *Ibid.*, 2:20 a–b.
14. Zhang Pan's gloss in the original text: "扐四揲之，後又加四揲，以求其小數，而决吉凶。如原之一一，則為元吉。 After four sortings, perform four more sortings to seek the 'small number' and determine good or bad luck. If it is 1.1 of [the Number Name] 'Origin', then it is 'Prodigious Good Luck.'" *Ibid.*, 2:20 a–b.

APPENDIX V

GLOSSARY OF *YIJING* HEXAGRAM AND COMMENTARY NAMES

H1 *Qian* 乾
H2 *Kun* 坤
H3 *Zhun* 屯
H4 Meng 蒙
H5 *Xu* 需
H6 *Song* 訟
H7 *Shi* 師
H8 *Bi* 比
H9 *Xiaochu* 小畜
H10 Lü 履
H11 *Tai* 泰
H12 *Pi* 否
H13 *Tongren* 同人
H14 *Dayou* 大有
H15 *Qian* 謙
H16 *Yu* 豫
H17 *Sui*隨
H18 Gu 蠱
H19 *Lin*臨
H20 *Guan* 觀
H21 *Shihe* 噬嗑
H22 *Bi* 賁
H23 *Bo* 剥
H24 *Fu* 復
H25 *Wuwang* 无妄
H26 *Dachu* 大畜
H27 *Yi*頤
H28 *Daguo* 大過
H29 *Kan* 坎
H30 *Li* 離
H31 *Xian* 咸
H32 *Heng* 恒
H33 *Dun* 遯
H34 *Dazhuang* 大壯
H35 *Jin* 晉
H36 Mingyi 明夷
H37 *Jiaren* 家人
H38 *Kui* 睽
H39 *Jian* 蹇
H40 *Xie* 解
H41 *Sun* 損
H42 *Yi* 益
H43 *Guai* 夬
H44 *Gou* 姤
H45 *Cui* 萃
H46 *Sheng* 升
H47 *Kun* 困
H48 *Jing* 井
H49 Ge 革
H50 *Ding* 鼎
H51 *Zhen* 震
H52 Gen 艮
H53 *Jian* 漸
H54 *Guimei* 歸妹
H55 *Feng* 豐
H56 Lü 旅
H57 *Xun* 巽

H58 *Dui* 兑
H59 *Huan* 渙
H60 *Jie* 節
H61 *Zhongfu* 中孚
H62 *Xiaoguo* 小過
H63 *Jiji* 既濟
H64 *Weiji* 未濟

First Order Commentaries (Ten Wings *shiyi* 十翼):
Tuan 彖
Xiang 象
Wenyan 文言
Xici shang 繫辭上
Xici xia 繫辭下
Shuogua 說卦
Xugua 序卦
Zagua 雜卦

APPENDIX VI

GLOSSARY OF CHINESE NAMES AND TERMS

bagua 八卦 eight trigrams
bayuan 八元 the "Eight Adepts" who served the legendary ruler Gao Xin
Cai Chen 蔡沈 (1167–1230)
Cai Yuanding 蔡元定 (1135–1198)
Cao Rong 曹溶 (1613—1685)
Chen Tuan 陳摶 (871?–989)
Cheng Yi 程伊 (1033–1107)
Cheng Zongshun 程宗舜 (fl. ca. 1550)
Daxue 大學 *Great Learning*, one of the *Sishu*
dayanzhishu 大衍之數 numbers of the great expansion
Dazai liji 大戴禮記 the Record of Rituals, Dazai version
Dao 道 the Way
dizhi 地支 12 earthly branches; along with *tiangan* form the duodecimal counting system
Dongji zhenjing 洞極真經
ershisi jieqi 二十四節氣 24 solar terms
Gao Xin 高辛 also called Digu 帝嚳, (3rd millenium BCE.?)
gua 卦 trigram(s) or hexagram(s)
Guan Lang 關朗 (+5th cen.?)

guaqi 卦氣圖 Array or the 64 hexagrams distributed over the 365 day solar year
Han Bangqi 韓邦奇 (1479–1556)
Hetu 河圖 [Yellow] River Chart
Hongfan 洪範 Master Plan, a chapter in the *Shujing*
Hongfanpai 洪範派 Hongfan School
huangzhong 黃鐘 middle C in traditional music theory
Huang Zongxi 黃宗羲 (1610–1695)
Jizi 箕子 (11th cen. BCE.)
Jiao Gong 焦贛 (1st cen. BCE.)
jiuchou 九疇 Nine Divisions of the *Shujing*'s *Hongfan* chapter
*jiujiu hangshu tu*九九行數圖 Nine by Nine Numbers in Rows Chart
junzi 君子 noble-minded person
Kong Yingda 孔穎達 (574–678)
Kongzi 孔子 Confucius (551–479 BCE.)
li 理 Pattern(s); sometimes translated Principle
Li Chunjin 李純謹 (fl. ca. 1518)
Li Dingzuo 李鼎祚 (ca. 8th cen.)
Liji 禮記 *Record of Rites*
Liu Mu 劉牧 (1011–1064)
Liu Shiqu 劉世衢 (18th cen.)
Lu Bian 盧辯 (ca. 520 CE)
Lunyu 論語 "Analects" of Confucius, one of the *Sishu*
Luoshu 洛書 Luo [River] Book
lülifa 履歷法 tonal ratio calendar
Mengzi 孟子 Mencius (372–289 BCE.), also title of one of the *Sishu*
nongli 農曆 agricultural calendar
ou 偶 even (number)
ou 耦 pair
Qi 氣 material substance; energy; the elements
qi 奇 odd (number)
qishierhou 七十二候 72 seasonal periods of year
Qianxu 潛虛
qushen 屈信 extension and intension
Sanwuli ji 三五曆紀 fragmentary 3rd cen. CE collection of cultural lore
Shao Yong 邵雍 (1011–1077)
shifa 筮法 divination method
Shujing 書經 *Book* [*of History*]
shuming 數名 Number Names

Shun 舜 (3rd Millenium BCE?)
Shuowen 說文 Chinese character dictionary (121 CE)
Sikuquanshu 四庫全書
Sima Guang 司馬光 (1019–1086)
Sishu 四書 the "Four Books" identified as essential texts by Zhu Xi
sixiong 四凶 the"Four Evil Ones" in the reign of Shun
suanchou jishufa 算籌記數法 counting rod format numbers
Taixuanjing 太玄經
tiangan 天干 ten heavenly stems; with *dizhi* forms the duodecimal counting system.
wanwu 萬物 the myriad things
Wang Qi 王圻 (1530–1615)
Wang Yinglin 王應麟 (1223–1296)
Wen Wang 文王 King Wen (11th cen., BCE.), founder of the Zhou Dynasty
wuchang 五常 Five Constants of correct behavior
- *li* 禮 ritual
- *ren* 仁 benevolence
- yi 義 morality
- *xin* 信 trustworthiness
- *zhi* 智 wisdom

Wu Wang 武王 King Wu (11th cen. BCE.), son of Wen Wang
wuxing 五行 Five Phases of phenomenal development
- *huo* 火 fire
- *jin* 金 metal
- *mu* 木 wood
- *shui* 水 water
- *tu* 土 earth

Xiangshupai 象數派 Image and Number School
xing 形 form
Xinglidaquan 性理大全, digest of Confucian philosophy published in 1415
Xiong Zongli 熊宗立 (1409–1482)
Xu Zheng 徐整 (3rd cen. CE)
yang 陽 the positive, aggressive, light natural force or developmental phase
Yang Hui 楊輝 (d. 1298 CE)
Yang Xiong 楊雄 (53 BCE.–18 CE)
yao 爻 individual "lines" of *Yijing* hexagrams
Yao Yong 姚鏞 (1191–?)

Yilin 易林 *Forest of Change*
Yongledadian 永樂大典, encyclopedia published in 1408
yin 陰 the negative, passive, dark natural force or developmental phase
Yu Chen 俞深 [*jinshi* 進士 1475]
Yueling 月令 "Lunar Seasons" chapter of *Liji*
zheng 正 righteousness
Zhanggong jun zhai 章貢郡齋, 13th cen. CE publishing house
Zhang Pan 章懋 (1437－1522)
Zheng Xuan 鄭玄 (127–200 CE)
Zhongyong 中庸 *Doctrine of the Mean*, one of the *Sishu*
Zhou Gong周公, the Duke of Zhou (11th cen., BCE.) regent for Wu Wang
Zhouyi jijie 周易集
Zhoui qimeng 周易啟蒙
Zhouyi zhengyi 周易正義
Zhu Xi 朱熹
Zhu Yizun 朱彝尊 (1629–1709)
zonghengtu 縱横圖 magic square

SELECTED BIBLIOGRAPHY

Andrews, William Symes. *Magic Squares and Cubes*. LaSalle, IL: Open Court Publishing Co., 1917; republished New York: Cosimo, 2004.

Ban Gu 班固. *Bohutong delun* 白虎通德論 https://ctext.org/bai-hu-tong/zh.

Cai Chen 蔡沈. *Hongfan huangji neipian* 洪範皇極內篇 [*Master Plan Supreme Pivot Inner Writings*]. https://archive.org/details/06065968.cn.

———. *Shu jizhuan* 書集傳 [*Collected Commentaries on the Book [of History]*]. https://ctext.org/wiki.pl?if=gb&res=731447.

Cai Xuesu 蔡學蘇, ed. *Writings of the Nine Confucians of the Cai Family* 《*Caishijiuru shu* 蔡氏九儒書》*Sanyu shuwu* 二餘書屋, 1868. (books.google.com)

Cammann, Schuyler. "The Evolution of Magic Squares in China," *Journal of the American Oriental Society*, 80 (2), April, 1960: 116–124.

Chan, Wing-tsit. *A Sourcebook in Chinese Philosophy*. Princeton, NJ: Princeton University Press, 1963.

Chen Shih-chuan. "How to Form a Hexagram and Consult the I Ching." *Journal of the American Oriental Society* 92, no. 2 (1972): 237–249.

Chen Zongshun 程宗舜. *Hongfan jianjie* 洪範淺解. https://archive.org/details/02094372.cn.

Chunqiu zuozhuan 春秋左傳 https://ctext.org/chun-qiu-zuo-zhuan/zh.

Dong Kai 董楷, *Zhouyi zhuanyi fulu* 周易傳義附錄. https://archive.org/details/06062816.cn.

Gushiciwang 古诗词网. https://www.gushici.com.

Han Bangqi 韓邦奇. *Hongfan tujie* 洪範圖解. http://www.360doc.com/content/20/0614/00/8541396_918340404.shtml.

Han Ying 韓嬰. *Han shi waizhuan* 韓詩外傳。 https://ctext.org/han-shi-wai-zhuan/zh.

Hanshu 漢書. https://ctext.org/han-shu/lv-li-zhi/zh.

Hu Fangping 胡方平, *Yixue qimeng tongshi* 易學啟蒙通釋 https://archive.org/details/06062832.cn.

Hu Guang 胡廣, et al. *Zhouyi zhuanyi daquan* 周易傳義大全 https://ctext.org/wiki.pl?if=gb&res=158704.

Huang Zongxi 黄宗羲, *Yixue xiangshulun* 易學象數論. https://archive.org/details/06063058.cn.

Kang Yu 康宇. "*Lun songyuan xiangshusichao xingqi ji qiduigudai shuxuefazhande yingxiang* 论宋元象数思潮兴起及其对古代数学发展的影响." *Ziranbianzhengfa yanjiu* 自然辩 证法研究, Vol. 34, No. 9:Sep., 2018: 93–99.

Lai Zhide 來知德. *Zhouyi jizhu* 周易集注 https://archive.org/details/06062891.cn.

———. *Zhouyi Liazhu tujie* 周易來注圖解. (Taibei: Huiwen chubanshe 惠文出版 社:1972.

Liji 禮記 [Record of Rites]. https://ctext.org/liji/zh.

Liu Xiang 劉向, ed. *Shuoyuan* 說苑。https://ctext.org/shuo-yuan/zh.

Lunyu 論語 [*The Analects [of Confucius]*]. https://ctext.org/analects.

Mengzi 孟子 [*Mencius*]. https://ctext.org/mengzi/zh.

Needham, Joseph, and Wang Ling, and Lu Gwei-djen. *Science and Civilization in China*. Cambridge: Cambridge University Press, 1959.

Nylan, Michael. *The Canon of Supreme Mystery by Yang Xiong*. Albany: State University of New York Press, 1993.

Ouyang Xun 歐陽詢。 *Yiwen leiju* 藝文類聚, https://ctext.org/yiwen-leiju/zh.

Pickover, Clifford. *The Zen of Magic Squares, Circles and Stars*. Princeton, NJ: Princeton University Press, 2002.

Schulz, Larry J. *Lai Chih-te (1525–1604) and the Phenomenology of the 'Classic of Change (I Ching).'* Princeton, 1982.

——— and Thomas J. Cunningham. "The Seasonal Structure Underlying the Arrangement of Hexagrams in the *Yijing*." *Journal of Chinese Philosophy* 17 (1990): 289–313.

———. "Structural Motifs in the Arrangement of the 64 *Gua* in the *Zhouyi*." *Journal of Chinese Philosophy* 17 (1990): 345–358.

———."Structural Elements in the *Zhou Yijing* Hexagram Sequence." *Journal of Chinese Philosophy* 38, no. 4 (December 2011): 639–665.

———. *Hexagrammatics*. Atlanta: Zizai, 2016.

———. *NGua Theory*. Atlanta: Zizai, 2018.

———, 舒來瑞. "*Zhouyi*" *guaxuwenti zonglun* 〈周易〉卦序问题综论. Wuhan: Wuhandaxue chubanshe 武汉大学出版社, 2022.

Sesiano, Jacques. *Magic Squares in the Tenth Century*. Cham, Switzerland: Springer, 2017.

Shaughnessy, Edward L. *I Ching: The Classic of Changes*. New York: Ballantyne Books, 1997.

Shijing 詩經 with James Legge's translation. https://ctext.org/book-of-poetry.

Sima Qian 司馬遷 *Shiji* 史記 (ctext.org) https://ctext.org/shiji/zh.

Smith, Richard J. "Fathoming the Changes: The Evolution of Some Technical Terms and Interpretive Strategies in Yijing Exegesis." *Journal of Chinese Philosophy* 40:S (2013): 146–170.

Songshi, 宋史. https://ctext.org/wiki.pl?if=gb&res=975976.

Sun Lesong 孫功松. "*Guodie yu guale—guo yonghe zhxi guanyu shifade bianlun* 過揲與掛扐— 郭雍和朱熹關於蓍法的辯論," https://blog.xuite.net/dejavu8899/blog/112300197

Tuo Yong 庹永. *Cai Yuanding fuzi yixuesixiang chanshi* 蔡元定父子易学思想阐释. Doctoral dissertation: Xiamen University 厦门大学, 2011.

Weisstein, Eric W. "Mathematics Notebook, Magic Squares" https://archive.lib.msu.edu/crcmath/math/math/m/m029.htm.

Wilhelm, Richard and Cary F. Baynes. *The I Ching*. Princeton: Princeton University Press, 1950.

Wu Cheng 吳澄. *Yueling qishierhou jijie* 月令七十二候集解 https://zh.wikisource.org/wiki/ 月令七十二候集解

Wu Wenjun 吴文俊, ed. *Zhongguo shuxueshi daji* 中国数学史大系, Vol. 5, *Liangsong* 两宋. Beijing: Beijing shifandaxue chubanshe 北京师范大学出版社, 2000.

Yang Hui 楊輝. *Xuguzhaiqi suanfa* 續古摘奇算法. https://ctext.org/wiki.pl?if=gb&res=139193.

Zhu Xi 朱熹. *Daxue zhangju* 大學章句。https://ctext.org/si-shu-zhang-ju-ji-zhu/da-xue-zhang-ju/zh.

———. *Zhongyong zhangju* 中庸章句。 https://ctext.org/si-shu-zhang-ju-ji-zhu/zhong-yong-zhang-ju/zh.

———. *Yuanben Zhouyibenyi* 原本周易本義 https://archive.org/details/06070838.cn.

———. *Zhuzi yulei* 朱子語類 https://ctext.org/zhuzi-yulei/zh.

——— and Joseph Adler. *The Original Meaning of the Yijing*. New York: Columbia University Press, 2020.

Zhuan hongguang 传洪光. *Fengshuiluopan quanjie* 风水罗盘全解. Beijing: Hualing chubanshe 华龄出版社, 2011.

INDEX

A

AA equivalence rule 21, 25, 31, 43, 138, 139
Abundance matrix 24
Achievement matrix 22
Adler, J.
Original Meaning of the Yijing, The 36, 38, 142, 170
Agricultural calendar 18, 43, 127, 142–150
in placement and diction 131–134
Andrews, W. S. 37
"knight move" method of construction 10
Annual solar cycle xvi, 2, 35, 42
Arabic numerals 15
Architecture
81 matrices xix, 28–29
of the images 138–139
verbal images 141
Associative and semi-magic squares 4, 10
Autumn Equinox 106

B

Base-nine numeric system 8
Book [of History] (Shujing) 1, 153
Bright virtue 136, 137

C

Cai Chen
Collected Commentaries on the Book [of History] (Shu jizhuan) 1
construction rules 25
divination method 29, 31, 37, 43, 127, 161, 163, 164
enlightenment, intelligence 136
Good and Bad Luck 31, 32
Hongfan huangji neipian xvii, 1, 140, 154, 156, 161
Image and Number tradition 2, 3, 6, 162, 175
"Inner Writings," 157

"Lamenting the Honorable Liu Wenan," 151
Luck Schedule 8, 9, 17, 18, 34, 41
Luoshu magic square xvii, 4–5
Master Plan 1, 2, 127, 161
matrices (verbal terms to numbers) xviii, 6
matrix construction 6, 35
tens complement rule 128
noble-minded person
fullness 141
Good Luck 46
qualities and discipline 135–138
vs. small-minded person 136
nonary system 4, 169
Number Names 6, 18
language of 127–128
set of 128
number theory 16, 139
numerical system 165
"Outer Writings," 156, 157
"Pattern," xviii
rationalized yarrow stalk divination method xvii
righteousness (*zheng*) 137
rules of construction 139
self-consistency 44
Songshi biography 153–154
stalk sorting methodology 6, 35
supreme pivot *(huangji)* 140
"the Numbers say," 18, 139, 161
visualization of numerical structures 141
work in intellectual history xix
Writings of the Nine Confucians of the Cai Family (Caishijiuru shu) 159
yang:yin 32
Yueling 139, 142
Caishijiuru shu 36, 43, 143–146, 148, 150, 159
Cai Yuanding (1135–1198) 1, 3, 35, 36, 153, 156, 173
Cammann, S. xvii, xviii
magic squares 4, 15
Evolution of Magic Squares 37
proposed proto-order 9 magic square 5, 6
Cao Rong 157, 173
Center and balance 41, 141
Center matrix 24
Cheng Yi (1033–1107) 37, 136, 151, 173
textual interpretations 1
Cheng Zongshun 6, 39, 157, 162, 173
Chen Tuan (871?–989) xvi, 6, 173
China
classical Chinese phrasing 43
line-by-line commentary 44
magic squares xvii, xx
mathematics xviii
names and terms 173–176
Number Names in Chinese 42
Circular and square matrices 3
Classic literature 134–135, 162
Column-to-column replacement strategy 10, 13, 26
Commentary names 171–172
Composite magic square 5
Confucian Classics xvii, 153, 162
Confucianism xvi, xx, 141
Constancy matrix 23
Cunningham, T. J.
"Seasonal Structure," 36, 150

D

Daoism xvi
Daxue 136, 137, 147, 148, 150, 151, 173
Divination method 1, 3, 29, 31, 32, 37, 41, 43, 127, 128, 161, 163, 164
bias toward positive outcomes 34–35
Dongji zhenjing 3, 36, 127, 154, 173
Double Good Luck term 33, 43, 138

E

81 Number Name Luck Spectrum matrices 25, 26

81 Number Name matrices
architecture and meaning 28–29
distribution of terms 29, 30
interrelationships 29
rules of construction 25–27
structure of 16–25
81 numbers, chart of
abundance *Shu* 5.1 82
achievement *Cheng* 2.1 55
agreement *Xi* 6.8 98
altruism *Gong* 4.4 76
anxiety *Ju* 7.2 101
arousal *Zhen* 2.3 57
ascension *Sheng* 5.4 85
bestowing *Xi* 4.8 80
center *Zhong* 5.5 86
change *Ge* 8.5 113
competing *Jing* 7.6 105
constancy *Chang* 2.5 59
contentment *Yu* 5.3 84
cunning *Ren* 9.5 122
danger *Li* 1.9 54
decision *Jue* 5.2 83
decoration *Shi* 6.1 91
decrease *Sun* 6.5 95
dimness *Mei* 6.4 94
dispute *Song* 7.8 107
distance *Yuan* 6.9 99
distinctness *Zhang* 4.6 78
dividing *Fen* 7.7 106
doubt *Yi* 5.8 89
emptiness *Xu* 6.3 93
end *Zhong* 9.9 126
excess *Guo* 5.7 88
extending *Shen* 1.4 49
falling *Duo* 9.8 125
firmness *Jian* 8.4 112
flowering *Hua* 2.9 63
following *Cong* 3.3 66
fruits *Shi* 8.1 109
fullness *Ying* 4.7 79
gladness *Xin* 3.8 71
guarding *Shou* 1.3 48
harvest *Shou* 7.9 108
hiding *Qian* 1.2 47
imprisonment *Qiu* 9.4 121
increase *Yi* 4.5 77
kinship *Qin* 2.8 62
knotting *Jie* 8.9 117
mating *Jiao* 3.4 67
meeting *Yu* 9.2 119
minimizing *Gua* 5.9 90
naiveté *Meng* 1.6 51
nearness *Bi* 4.1 73
nourishing *Yang* 9.1 118
nurturing *Yu* 3.5 68
opening *Kai* 4.2 74
origin *Yuan* 1.1 46
peace *Chong* 2.2 56
peril *Wei* 8.3 111
perversity *Li* 6.2 92
prayer *Qi* 2.4 58
progress *Jin* 4.3 75
quickness *Xun* 7.1 100
repayment *Bao* 8.6 114
repose *Bi* 1.7 52
revelry *Mi* 4.9 81
ridding *Chu* 7.3 102
rigidity *Gu* 9.6 123
seeing *Jian* 3.1 64
shifting *Yi* 9.7 124
sickness *Ji* 7.5 104
softness *Rou* 2.6 60
sojourn *Bin* 8.2 110
soothing *Shu* 3.9 72
stopping *Zhi* 8.7 115
straightness *Zhi* 1.5 50
strength *Zhuang* 3.6 69
subservience *Fu* 5.6 87
taking *Huo* 3.2 65
tranquility *Yi* 2.7 61
triumph *Sheng* 9.3 120
uplifting *Xing* 3.7 70
utility *Yong* 6.6 96
waiting *Xu* 1.8 53
weakness *Ruo* 7.4 103
weapons *Rong* 8.8 116
withdrawal *Shi* 6.7 97

English
Number Names in 42, 43
Equivalence rule 21, 25, 31, 43, 138, 139

F

"Fair" divination system 32–35, 135
Fengshui xvi, 38, 159
Five Phase *(wuxing)* theory xvi, xx, 2, 134, 144, 149, 158

G

Graphic depictions and divination xv
Graphics
binary figures xvii
concrete images xviii
hexagrams xvi, xvii, xviii
Image and Number proponents xv
visualizations of numerical relationships xv
Gregorian calendar 142–150
Guan Lang 3
Dongji zhenjing 127

H

Han Bangqi (1479–1555) 37, 157, 159, 174
graphics and divination methods 162
Han Dynasty xvi, xix, 135, 142, 162
Hansh 3, 6, 135, 144, 158
Headings *(shou)* 127
aspect conversion 167
Hetu xv, 36, 158, 174
Hexagrams xv, xvi, xvii, xviii, 31, 135
Guaqitu 3
H4 *Meng* 44, 130, 133, 143, 150
Hiding matrix 20
Hongfan xvii, 1, 154, 156, 157
divination method 163–169
Nine Divisions *(jiuchou)* 2, 4
Hongfan neipian xx, 157
Hongfan School *(Hongfanpai)* xviii, 161–162
Huangji jingshi shu xx, 157
Huang Zongxi 40
Yixue xiangshulun 161, 163
Hybrid theory xvii

I

Images
allusions to the *Yijing* 134–135
architecture of 138–139
chart of the 81 numbers, 45–126
classic literature 134–135
Image and Number School xv, xix, 3
noble-minded person
qualities and discipline 135–138
number names
agricultural calendar in the placement and diction 131–134
language of Cai Chen's 127–128
set of 128–131
verbal images xi, xvi, xviii, 38, 42, 141

J

Jiao Gong 174
Forest of Change (Yilin) 127
Jing Fang
guaqitu 2

K

Kang Yu
"waves" of Image and Number xix, 3, 6, 37
Kongzi xvi, 136, 148, 149, 162

L

Lai Zhide (1525–1604) xv
  Zhouyi 162
Language of Number Names 127–128
Liji 18, 43, 135, 145
Lines *(yao)* xviii, 134
Lu Bian 158, 174
Luck Schedule 2, 25
  Column Replacement Pattern 11
  Column Replacement Strategy 12
  replacement scheme 19–21
Luck Schedule matrix 9, 13, 19, 21, 23–25, 27
Luck Schedule terms
  balance 140
  error 40
  frequency of 33
  in alpha and numeric 8
  in Number Name matrices 19
  positive and negative qualities 32
  probabilities of occurrence 29–34
Luck Spectrum terms 139
Lunar calendar xvi, 2, 18, 43, 142, 176
Lunar Seasons *(Yueling)* 18, 43, 134, 139, 142, 144–146, 148, 149, 150, 176
"Luo [River] Book," 2
Luoshu xv, xvii, 2, 25, 26, 139
  "Chart of Eight Numbers in Opposition *(Bashuxiangduitu)* 12
  configuration 4
  magic square 4–5
  replacement formula 13
  "tens complement" rule 4

M

Magic square theory 10
  Cammann's Proposed Proto-Order 9, 5
  history of mathematics xviii
  in China xvii, xx, 37
  Luoshu 4–7
  Order 9, 5, 15
Mao xvi
"Master Plan *(Hongfan)*" xvii, 4, 156
Matrix
  abundance matrix 24
  achievement matrix 22
  center matrix 24
  complement rule in matrix construction 128
  complex suite of 141
  constancy matrix 22
  construction 35
  81 Number Name matrices 2, 16–19
  elements of 17
  hiding matrix 20
  internal numerical structure 139
  interrelationships of the 81 matrices, 29
  Luck Schedule matrix *See* Luck Schedule matrix
  Numeric Luck Schedule Matrix 9
  origin matrix 19
  square matrices 3, 14
  straightness matrix 21
Meng Xi
  Guaqitu 127
  guaqitu 2
Multi-layered imagistic structure 138

N

Name Spectrums 30, 33
Needham, J.
  Science and Civilization in China xviii
NeoConfucianism 3, 141
NeoConfucians xviii, 2, 35
NewBook of Pitch Pipes, The 3
Nine by Nine Circular Number Graph 14
Nine by Nine graphics 42
9 x 9 Number Name matrix 25, 26
Nine by Nine Numbers in Rows Chart 6, 15, 19
"Nine palace" method 158
Noble-minded person
  fullness 141

Good Luck 46
Pattern 46
qualities and discipline 135–138
vs. small-minded person 136
Nonary system 4, 169
Number(s)
and number sets 6–14
defined 6, 41
81 matrix suite
architecture and meaning 28–29
rules of construction 25–27
structure of 19–25
81 Number Name matrices
structure of 16–19
luck schedule terms
probabilities of occurrence 29–34
Luoshu magic square 4–5
square matrix
cyclical template of 81 nodes, 14–16
12th-century divination systems 34–35
Number Names
agricultural calendar
in placement and diction 131–134
antonyms among 129
in Chinese 42
in English 42
language of Cai Chen's 127–128
opposed across seasonal continuum 132
set of 128–131
verbal architecture 138
Numerical relationships, visualizations of xv
Number sets 6–14
Number theory 1, 16, 139
Numeric Luck Schedule Matrix 9
Numerical system 165

O

Odd-even placement rule 10
Origin matrix 20

P

Probabilities of occurrence 23
Luck Schedule terms 29–34
Profit in difficult propriety 138

Q

Qianxu 3, 36, 127, 141, 154, 158, 161, 174
Qing Dynasty 43

R

Rationalized yarrow stalk divination method xvii
Record of Rites (Liji) 18, 43
Rule-based semi-magic square 141

S

Semi-magic square 4, 15, 22, 141
"Chart of the Eight Numbers," 10
Set of Number Names 128–131
"72 seasonal periods *(qishierhou)*" 18, 43, 174
Shao Yong (1011–1077)
8 x 8 square hexagram matrix xvi, xviii
Huangji jingshi 3
number theory 1
Shiji 135, 144, 146, 149
"Biographies of Astrologers," 158
Shijing 1, 135, 150 *See also Hongfan*
Shu (Number) 1–40
Shujing xvii, 37, 38, 135, 140, 144, 145, 147
Hongfan 4, 148, 157, 174
Shuogua xvi, 157, 172
Sikuquanshu xx, 4, 6, 36, 39, 43, 155, 159, 160–163

Sima Guang 3
  Qianxu 127
  Xuqian 38
Smith, R. 150, 162
Solar calendar 2, 15, 18, 19, 35, 36, 41–43, 85, 131, 139, 142–150, 173, 174
Spring Equinox 65, 66
Square matrix
  cyclical template of 81 nodes, 14–16
Stalk sorting 29
Straightness matrix 21
Summer Solstice 85, 86, 131, 133
Sun Lesong 39
Supple grass 80, 134, 146
Supreme Pivot *(huangji)* xvii

T

Taichuli 3, 142
Tens complement rule 10, 11
Tetragrams 2, 3, 127
Tonal ratio calendar 3, 174
Trends of luck 23
Trigrams xv, xvi, xviii, xix, 2, 7, 37, 38, 143, 144, 154, 158, 166, 173
  Taixuanjing 3
Tuan 134, 142, 143
12th-century divination systems
  bias toward positive outcomes 34–35

W

Wang Qi (1530–1615)
  Xuwenxian tongkao 157
Wang Yinglin (1223–1296)
  Master Plan Numbers 157
Weisstein, E. W. 37
Wenming 134, 144, 145
Wenyan 38, 134, 142, 143, 145, 149, 172
Winter Solstice xvi, xviii, 15, 19, 38, 46, 131, 133
Writings of the Nine Confucians of the Cai Family (Caishijiuru shu) 43, 159
Wu Cheng
  Yueling 144–146, 148, 149
Wu Wenjun xviii, xx

X

Xiang xviii, 2, 41–151
Xici 2, 131, 134
Xingli Daquan 156, 157
Xiong Zongli (1409–1482) 157, 161, 163
  definitive commentary 6

Y

Yang Hui (d. 1298 CE)
  magic squares 5, 15
  vertical and horizontal charts xvii, xviii
  Xuguzhaiqi suanfa 4, 5
Yang Xiong (53 BCE–18 CE)
  Taixuanjing 2, 127
  tetragrams 127
Yao Yong 18
  Sikuquanshu 159
Yarrow stalk-sorting divination method xv, 34
Yijing 6
  allusions to 134–135
  divination 134, 164
  graphics
    binary figures xvii
    concrete images xviii
    hexagrams xvi, xvii, xviii
    Image and Number proponents xv
    visualizations of numerical relationships xv
  hexagrams 34, 44, 134, 171–172
Yijing Xici 7, 34, 37, 142, 145, 148, 154
Yongle Dadian 156, 157
Yu Chen 176

Master Plan Divisions Explained (Hongfan choujie), The 157

Z

Zhanggong jun zhai 159
Zhang Pan (1437–1522) 163, 165, 170
Zhen xvi, 37, 57
Zhouyi Benyi 43
Zhu Xi (1130–1200) xvii, 1, 34, 159
Book [of History] 153
graphic depictions and divination xv
NeoConfucian ontology 2, 141
Original Meaning of the Yijing, The 36
Shujing 135
stalk sorting concept 34, 165, 169
works on Confucian texts 1
yarrow stalk-sorting divination method xv
Zhongyong 141
Zhu Yizun (1629–1709) 176
Jingyi kao 157

www.ingramcontent.com/pod-product-compliance
Ingram Content Group UK Ltd.
Pitfield, Milton Keynes, MK11 3LW, UK
UKHW020841290926
15470UKWH00004B/1482